BIOLOGY
DISCOVERY
ACTIVITIES KIT

BIOLOGY DISCOVERY ACTIVITIES KIT

Lessons, Labs, and Worksheets for Secondary Students

MARY LOUISE BELLAMY

Illustrated by
Karen Teramura

THE CENTER FOR APPLIED
RESEARCH IN EDUCATION
West Nyack, New York 10995

10   9   8   7   6   5   4   3   2   1

Excerpts on procedures for Biuret, Benedict's, and Sudan III
tests adapted from *A Sourcebook for the Biological Sciences, third
edition,* by Evelyn Morholt and Paul F. Brandwein, copyright
1986 by Harcourt Brace Jovanovich, Inc. Reprinted by permission
of the publisher.

Library of Congress Cataloging-in-Publication Data

Bellamy, Mary Louise, 1955–
    Biology discovery activities kit : lessons, labs, and worksheets
for secondary students / Mary Louise Bellamy : illustrated by Karen
Teramura.
        p.   cm.
    ISBN 0–87628–186–2
    1. Biology—Study and teaching—Activity programs.   I. Teramura,
Karen.   II. Title.
QH316.4.B44   1991
574′.0712—dc20   90–20050
                                                                              CIP

ISBN   0-87628-186-2

THE CENTER FOR APPLIED
RESEARCH IN EDUCATION
BUSINESS & PROFESSIONAL DIVISION
A division of Simon & Schuster
West Nyack, New York 10995

Printed in the United States of America

To Randy

Acknowledgments

I would like to thank the following people:

Nicholas Sliz, for his contributions to the Genetics section;

Dr. J. David Lockard and Dr. Linda Sanders, for their valuable advice and technical assistance;

Dr. Elizabeth McMahan, for teaching me the beetle heart experiment used in "Health Store Products—Do They Work? You Find Out!";

Karen Teramura, for the beautiful illustrations and help in developing them;

my editor, Sandra Hutchison, for her help, patience, and support;

Zsuzsa Neff, my production editor, for her help, kindness and patience;

the biology teachers I have had as colleagues, from whom I have learned so much; and

my family, for their support and encouragement.

About the Author

Mary Louise Bellamy received her doctorate in science education from The University of Maryland at College Park. She earned a B.A. in zoology from The University of North Carolina at Chapel Hill, and an M.A.T. in science education from Emory University in Atlanta.

She has taught high school biology and biochemistry at public and private schools, including the Governor's School of North Carolina (Winston-Salem) and the North Carolina School of Science and Mathematics. Her students have won state and national science competitions, including the NASA/NSTA Student Shuttle Involvement Project. She has given presentations to teachers at national and state conventions.

Mary Louise was a North Carolina Outstanding Science Teacher in 1986, and won a Growth Initiatives for Teachers (GIFT) grant from GTE Corporation in 1984.

About This Resource

During the seven years I have taught high school biology, it has been my experience that we develop some of our best ideas and insights about teaching when discussing classroom strategies with other teachers. My purpose in writing *Biology Discovery Activities Kit* is to share with you ideas I have found successful in encouraging active student participation in learning—and to do so in a format that makes these ideas easy for you to use.

This book is designed to supplement traditional biology texts and laboratory manuals, not to replace them. The activities span a variety of topics and lend themselves to integration into many subjects in secondary biology courses.

I wrote the activities with teachers of all levels in mind and have tried to be particularly sensitive to the needs of beginning teachers. I remember vividly how difficult my first two years of teaching were and want very much to help make teaching easier for others who are beginning their careers. I also want all teachers to be able to use some of the strategies contained here as inspiration for creating other activities of their own.

The activities included vary in type and subject matter, but with one common thread woven through all of them: that students learn most effectively and most enjoy learning when they are active participants, thinkers, and creators of their own learning experiences—with guidance from a good teacher.

We are preparing students to live in a future which we may be unable to predict or comprehend. Since we cannot know what problems they will face, it is more effective to help students learn how to think than to teach them only facts. This is not to say that facts are not important; they are absolutely necessary, particularly in the study of biology. But if facts are all we teach, there are problems:

1. *Facts change rapidly in science.* Remember your own high school biology course? Unless it was taught quite recently, the text used in that course inevitably contained passages that were not merely misleading or incomplete, but—as measured by current knowledge—totally wrong. Yet these "facts" were perhaps conveyed to you as unchanging, immutable truth!

2. *Students must experience the process of science,* not merely learn about its results. Students experience exhilaration as they figure things out for themselves. In learning how "to science," they are learning how to think and solve problems, essential skills in science and in all other fields of knowledge.

The activities in this book encourage the development of active thinking by students in a variety of ways. Many allow students to experience or simulate one or more processes engaged in by scientists. For example, in many of the labs (such as "What Is the Effect of Radiation on Plants?"), students must form and test hypotheses by designing their own experiments. Other activities (such as "How Are Traits Transmitted in Corn Plants?", "How Do Mutations Occur?", and "What Is a Food Pyramid?") require students to deduce principles solely through observation or simulation of biological processes. The games "What Are Relationships Between Cell Structures and Photosynthesis?" and "What Are Relationships Between Cell Structures and Cellular Respiration?" require that students draw new relationships between concepts in biology.

The game entitled "What If There Were No Fungi?" asks students to predict the consequences of a hypothetical situation (that there are no fungi on earth) and propose a solution.

Activities such as "How Will You Solve the Plant Cell's Problems?" and "How Is the Cell Membrane Held Together?" ask students to "invent" a solution to a problem nature has solved already. The student invention is then compared with the "actual" solution. Others (such as "Can You Design a New Life Form?") require students to use their knowledge of biology to solve hypothetical problems.

Science is used to propose solutions to real-world problems that at present have not been solved in the activity "Health Store Products—Do They Work? You Find Out!" in which students test the claims of health products.

All of us, including scientists, increasingly must consider the implications of scientific knowledge to the rest of society—and the decisions these implications force us to make about real-life ethical dilemmas. In all of the activities in the section called "Controversial Issues in Biology" students gain practice in this type of decision-making.

Using This Book

For each activity, you will find detailed directions for preparation and teaching, a list of goals, suggestions for evaluating student performance, and reproducible student worksheets. A list of *Relevant Subjects* suggests areas of the biology curriculum into which the activity can be integrated. *Age/Ability Levels* described should be used as a general guideline of the difficulty of the activity. Since you are the best judge of the capabilities of your students, you should read each activity thoroughly before deciding whether it is appropriate for them. The *Entry Skills* listed will help you make this determination; make certain your students have mastered them before asking them to attempt each activity, since these are not taught as part of the exercise.

In most of the experiments students are asked to design themselves, the procedure they are to use has been specified in detail. Students at these age levels typically lack the necessary skills to design their own experiments entirely "from scratch." Forcing them to attempt such a task can create enormous frustration for both students and teachers. Building some structure into the activity and limiting the number of materials available to students means that the labs—while encouraging active student involvement—are nonetheless manageable for teachers.

In many of the lab activities, I have attempted to modify traditional labs to make them less "cookbook" in nature, to require more thinking and investigation on the part of the students. Because many of the lab techniques are commonly used, a minimum of new preparation on your part should be required.

You may photocopy the student worksheets and other materials provided as often as necessary to implement the activities. Enjoy!

Mary Louise Bellamy

A Note About Safety

Although I have tried to list safety precautions unique to each activity, there is not room to include all safety guidelines a teacher should be aware of at all times. For example, it is assumed that all teachers will have safety equipment such as a fire blanket, an eye wash station, and a first aid kit available at all times. It is assumed that students will be told to wear proper clothing, such as shoes that cover their entire feet instead of sandals, and sleeves that are not likely to be caught accidentally in a laboratory flame. Consult a local or state safety manual if you need a list of these general safety guidelines.

For some of the activities involving bacteria, I have warned against incubating them at 37° C, because pathogens grow at that temperature. However, in "Is the Water Around You Safe to Drink?" it is recommended that bacteria be grown at this temperature in order to obtain the number of *E. coli* needed to perform the experiments. Safety precautions are given in these activities for handling cultures grown at 37° C.

(Many of the safety precautions in this book are from the *Maryland School Science Safety Project: Safety Manual K–12,* © Maryland State Department of Education, 200 W. Baltimore St., Baltimore, MD 21201; used by permission.)

Contents

section I

Cell Structure and Chemistry

This section represents successful attempts to "breathe life" into units on cell structure and cell chemistry. In my experience, these topics are especially difficult to teach in a way that encourages active student involvement—in part since most cells are so small and their contents are not easily manipulated.

The first two classroom activities ("How Will You Solve the Plant Cell's Problems?" and "How Is the Cell Membrane Held Together?") both require that students propose solutions to "problems" already solved by nature. Concepts are always more interesting if a student attempts to figure them out for himself or herself first. Ideas presented in this way are also more likely to be remembered.

The other three activities are labs requiring that students form hypotheses and design experiments to test them. As with most of the experiments presented in this book, students are guided through experimental design in these labs by having basic procedures described in detail. If your students are advanced, they might be able to design their own experiments without these procedures.

An unusual or "discrepant" event is presented to students at the beginning of two labs that should pique their interest. One is the change in color of an extract of purple cabbage ("Why Does a Plant Pigment Change Color?"), and the other is a tube of gelatin which has failed to solidify after overnight refrigeration ("How Will You Determine the Effect of Heat on the Activity of an Enzyme?").

How Should These Activities Be Used?

Your imagination should be your guide as you integrate these activities into your biology units. Here are a few suggestions, however:

1. Introduce a unit on cell structure with "How Will You Solve the Plant Cell's Problems?" Students should then be able to understand the structures of cells in terms of their functional value (as solutions to problems) to the plant.

2. "Why Does a Plant Pigment Change Color?" is an excellent way to introduce processes of the scientific method in studying cells, since it is a relatively simple experiment and the results are readily visible.

3. "How Is the Cell Membrane Held Together?" may precede a lesson on active transport and follow one on the properties of water and its role in maintaining life.

4. "How Will You Determine the Effect of Heat on the Activity of an Enzyme?" is a great follow-up to lessons on enzyme action or protein structure. Students are more likely to remember the concepts after they complete this lab, since it involves gelatin and pineapple, both familiar foods.

ACTIVITY 1

HOW WILL YOU SOLVE THE PLANT CELL'S "PROBLEMS"?

Goals

After completing this activity, students will:

1. be familiar with some of the structures of plant cells; and
2. understand how the structures of plant cells relate to functions of the plant.

Synopsis

Certain "problems" nature has solved for plants are presented to students as unsolved problems. Students compare their solutions to these problems with the plant structures which represent "nature's answer."

Relevant Topics

cellular structure and function
photosynthesis

Age/Ability Levels

grades 7–10, most ability levels

Entry Skills and Knowledge

Materials

student worksheets: "Solving the Plant Cell's 'Problems'"
models or diagrams of the plant cell and subcellular structures

optional:
drawing pencils
colored pencils
evaluation worksheet

DIRECTIONS FOR TEACHERS

Preparation

1. photocopy student worksheets
2. obtain diagrams or models of plant cells and their subcellular structures
3. assign students to bring in drawing or colored pencils, if they are to be used

Teaching the Activity

1. Distribute student worksheets.
2. Go through an example with students to be sure they understand the assignment. For instance, a possible response to problem 1 (refer to student worksheet) might show a plant propping itself up on stilts. A solution to problem 6 might be that the flowers of the plant have a well marked "runway" on which insects can "land" easily. (see figure 1–1).

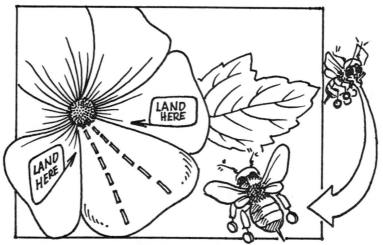

FIGURE 1–1. SAMPLE SOLUTION TO PROBLEM 6

NOTE: This activity assumes no prior knowledge of cell structure, so make sure students know that they are not being graded on whether their answers are right or wrong, but on their effort and imagination.

3. It is your decision:
   - whether to allow students to work individually or in groups of two; and
   - how many of the problems to assign—a reasonable number is two or three problems per group of students in a twenty-minute period.

   Be sure at least one student is working on each problem.

4. Provide encouragement and clarification about the assignment as students work, but the ideas should come from them.

   NOTE: For problem 2, it is true that some plants are carnivorous. These plants undergo photosynthesis, nonetheless.

5. Reassemble students as a class and call for volunteers to draw or describe their solutions. Accept all reasonable answers. Keep a record on the board of suggestions.

6. Show students the "real" solutions by using cell models, diagrams or both. As you do, however, discuss the plausibility of some student responses that do not represent "correct" answers. Point out that, had living things developed differently on this planet, some of the student solutions might have been found in nature. Some "real" solutions to the problems given include these:

Problem 1 Solutions:
- Hard substances (such as cork and lignin) are secreted by cells of woody plants.
- Turgor pressure is created by vacuoles which squeeze against cell walls; these are fortified to withstand this pressure.
- A plant wraps itself around a fence or another woody plant for support.

Problem 2 Solution:
- The chloroplast converts energy of the sun into sugars used by cells (by photosynthesis).

Problem 3 Solution:

- Water passes through nonwaterproof cells in root hairs.

Problem 4 Solution:

- Small channels called *plasmodesmata* connect adjacent cells.

Problem 5 Solution:

- These gases pass through stomata.

Problem 6 Solutions:

- Brightly colored pigments, found in chromoplasts of flower and fruit cells, attract insects and birds.
- Chemicals are made which attract insects and birds.

Problem 7 Solutions:

- Certain cells secrete a waterproof covering to prevent water loss from plant surfaces.
- Stomata close during the hottest part of the day, when evaporation is greatest.

Problem 8 Solution:

- Sugars produced by the plant cell (see solution to problem 2) are converted into cellular energy.

Problem 9 Solution:

- The nucleus, containing the "blueprint" of the cell in DNA, acts as the coordinator of cellular activities.

Problem 10 Solution:

- The Golgi apparatus acts as the packaging station, modifying material to be exported and keeping it separate from the rest of the cell.

Problem 11 Solution:

- The cell makes duplicate copies of subcellular structures, then, when large enough, pinches in half to make two cells.

Problem 12 Solutions:

- A strong external cell wall helps strengthen cells.
- An internal "cytoskeleton" made of a network of proteins acts as an internal support system.

Follow-up Activities

- Bake or buy a large rectangular cookie (this is the shape of a typical plant cell). With decorative icing in different colors, have students "create" an edible cell. Without referring to notes, each volunteer must:

—tell the class which part of the cell he or she is drawing;

—explain the function of the structure; and

—draw it in the appropriate region of the "cell."

When the cell is complete, allow students to enjoy their creation.

• Have students solve these "problems" for the cells of certain animals:

—the animal must be able to camouflage itself to prevent predation;

—it must be able to move about to seek food;

—it must be able to attract a mate of the opposite sex;

—an animal living in a dry climate must conserve water; and

—an animal living in a cold climate must keep itself warm.

• For each operation in Table 1-1 performed in a large country, have students write the structures in plant or animal cells which are responsible for analogous functions.

Table 1–1

Function in a Country (With Location)	Analogous Cell Structure (ANSWERS)
1. Control and organization (government)	nucleus
2. energy production (power plants)	mitochondria
3. communication system (telephone lines)	plasmodesmata
4. control of immigration and emigration (at the border)	cell membrane, cell wall
5. site of business transactions (office buildings)	cytoplasm, enzymes
6. food production (farms)	chloroplast
7. preparation of goods for export (packaging stations)	Golgi apparatus
8. creation of new goods (factories)	ribosomes

Evaluation

Method Evaluation	Goal(s) Tested
1. Completion of student worksheet	1, 2
2. Participation in class discussion	1, 2
3. Evaluation worksheet	1, 2

Answers to evaluation worksheet:

1. cell membrane
2. chromoplast
3. ribosomes
4. chloroplast
5. Golgi complex

Name _____   Date _____

SOLVING THE PLANT CELL'S "PROBLEMS"

Plants have developed so that certain "problems," described below, have been solved by their cells. Some of these are unique to plants; others are common to animals, as well.

Your Task

Your teacher will assign you one or more of the "problems" listed below which has been solved by plants. For each, indicate how a plant cell or group of cells might solve the problem. You may do this by making a drawing, by describing your solution in writing, or both. Afterwards, your teacher will help you compare your ideas to actual solutions found in plants.

Problem 1:   The plant must be able to "stand up," since it has no bony skeleton as many animals do.

Problem 2:   The plant must manufacture its own food, since it cannot eat in the same way many animals do.

Problem 3:   The plant must be able to obtain water from the environment.

Problem 4:   This cell must be able to communicate with other cells in the same plant.

Problem 5:   The plant must be able to obtain carbon dioxide from the outside world and rid itself of oxygen.

Problem 6:   The plant must be able to attract birds and insects, which assist it in reproduction.

Problem 7:   A plant which lives in a very dry climate must conserve its water.

Problem 8:   The plant must have energy to power all of its functions.

Problem 9:   The cell needs a "central planning unit" which will coordinate all its activities.

Problem 10:   The cell must have a way of preparing materials for export.

Problem 11:   The plant must be able to grow.

Problem 12:   The cell must be able to prevent itself from collapsing under the weight of other cells.

Name _____ Date _____

SOLVING THE PLANT CELL'S "PROBLEMS"
EVALUATION QUESTIONS

Different experiments were done on plant cells to remove or block the synthesis of one of the cellular components. The result of each experiment is briefly described below. From the description, in the space provided, write in the name of the cellular component that was most likely affected. Your answers will come from the following list of cellular structures:

ribosomes

nucleus

leucoplast (amyloplast)

chromoplast

mitochondria

chloroplast

endoplasmic reticulum

cell membrane

cytoplasm

Golgi complex

Results of experiments:

_____1. The entire contents of the cell spills out into its surroundings.

_____2. The cell loses its bright orange color.

_____3. Messenger RNA can be synthesized, but the remaining steps in protein synthesis cannot take place.

_____4. The cell loses its capacity to harness the sun's energy to manufacture sugars.

_____5. Secretions cannot be packaged for export to other cells.

ACTIVITY 2

HOW IS THE CELL MEMBRANE HELD TOGETHER?

Goals

After completing this activity, students will:

1. be familiar with the biochemical makeup of the molecules of the cell membrane;
2. understand the biochemical reasons for the structure of the cell membrane; and
3. know how molecules making up the cell membrane are arranged.

Synopsis

Students are guided through the process of "inventing" the phospholipid structure of the cell membrane.

Relevant Topics

cellular structure
biological molecules

Age/Ability Levels

grades 10–12, average and above average

Entry Skills and Knowledge

Before participating in this activity, students should be able to:

1. define the term "cell"; and
2. state a function of the cell membrane

Materials

student worksheets: "The Cell Membrane," "Cell Membrane in Cross Section," and "Phospholipid Molecules"
scissors
glue
cell transparencies 1 and 2

optional:
commercial model of a cell, showing cross sectional structure
electron micrograph of a cell in cross section
evaluation worksheets

DIRECTIONS FOR TEACHERS

Background on Cell Membrane Structure

The cell membrane is arranged as a phospholipid bilayer with the polar ("water-loving") ends of the molecules arranged end-to-end (see Fig. 2–1). The polar "heads" face the interior and exterior surfaces; the nonpolar ("water-hating") "tails" face each other on the interior of the cell membrane. Since all polar regions of the molecules are in contact with polar surfaces, and the nonpolar regions are in contact with each other, the solubility rule is satisfied. The solubility rule dictates that "like dissolves like"—a polar substance will dissolve another

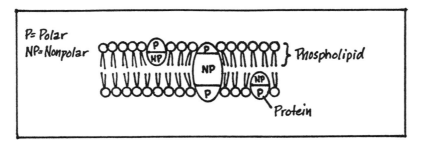

FIGURE 2–1 DISPERSAL OF PROTEINS AMONG PHOSPHOLIPIDS

polar one, and a nonpolar substance will dissolve another nonpolar one. Proteins, interspersed among the phospholipids, generally carry the same charges as the regions of phospholipid molecules around them, as shown in Figure 2-1.

Preparation

1. photocopy student worksheets.
2. obtain glue and scissors.

Teaching the Activity

1. Be sure students understand what cells are, and at least one function of cell membranes.
2. Distribute student worksheets, glue, and scissors, and briefly explain the assignment. It is your decision whether or not to allow students to work in groups of two, or individually. Remind students to exercise caution when handling scissors.
3. As students arrange phospholipid "molecules" on their worksheets, write your "OK" in the teacher approval space before they begin gluing them down. Provide encouragement as needed, but try to force students to find the correct arrangement themselves. Make a note of incorrect ones to discuss later. The correct arrangement, when finished, will look like the diagram shown in Figure 2–2.

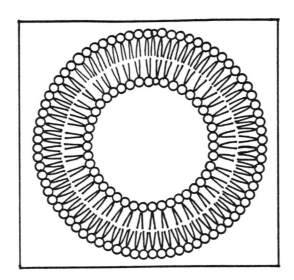

FIGURE 2–2. CORRECT PHOSPHOLIPID ARRANGEMENT

4. Assemble the students as a class, calling for volunteers to draw their structures on the board.

5. Explain the "solubility rule" to students, and give them the definitions of "polar" and "nonpolar" molecules. Have them label the regions of their diagrams which are polar, and those which are nonpolar. Then ask them to explain how the structure of the cell membrane satisfies the solubility rule.

6. Draw two or three of the incorrect arrangements you saw as you observed student work, and ask students why each would be impossible, according to the solubility rule. Some possible incorrect arrangements you may see include those represented in Fig. 2–3.

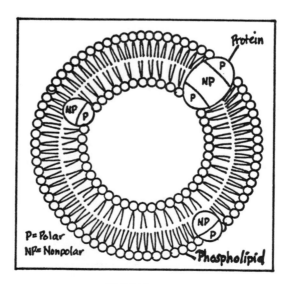

FIGURE 2–3.

7. Using a chalkboard drawing or transparencies (made from diagrams on the following pages) of the cell membrane, illustrate that in addition to phospholipid molecules, the cell membrane also contains many proteins. Have students draw some proteins on their diagrams as they observe those in the transparencies. Then have them label the different regions of the proteins as to whether they are more likely to be polar or nonpolar. A completed drawing may look like Fig. 2–4.

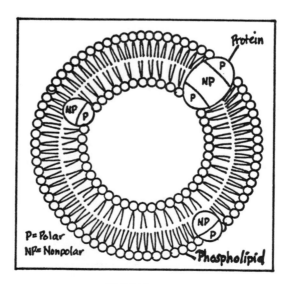

FIGURE 2–4.

8. Be sure students understand that their diagrams are not drawn to scale and that cell membranes are much thinner in relation to cell area than they appear on the drawings. You can illustrate this by showing students a model or electron micrograph of a cell and pointing out the membrane region.

9. (Optional) Ask the following questions to stimulate discussion. (Answers are given in parentheses after each question.)

   ● How do polar molecules pass through the cell membrane? (through polar "channels" in the proteins)
   ● Why do substances need to move across the cell membrane? (so that the cell can obtain nutrients and eliminate wastes)
   ● What substances are found inside the cell besides water? (proteins, including enzymes, cellular organelles) What might be some of their functions? (cellular respiration, control of heredity, manufacture of sugars)
   ● What substances might be found outside the cell, besides water? (proteins, other cells, food molecules, bacteria)
   ● What functions might the cell membrane have, other than those discussed here? (a semipermeable membrane barrier to the outside, immune functions, protection of the cell)

Follow-up Activities

The first two follow-up activities represent other examples of processes involving the "solubility rule."

● Have students research the chemical forces involved in detergent action. Detergents form micelles, structures which are created according to the same rules as cell membranes. Most dirt is nonpolar, and it becomes trapped in the middle of the micelles, as shown in Fig. 2–5.

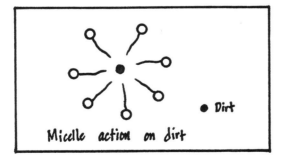

FIGURE 2–5. MICELLE ACTION ON DIRT

Since the outside is polar, water can carry away the dirt, now "trapped" in the micelle. (*NOTE*: You may choose to present the detergent example to students as a problem to be solved, in much the same way they solved the cell membrane structure.)

- Have students research how bile salts work to emulsify fats. Their action and structure are similar to those of micelles. Bile salts emulsify fats so that they can be transported in soluble form in the blood.
- (For advanced students:) "Invent" a way for the cell membrane to allow certain molecules into the cell and to exclude others. State one reason for your model being feasible and one reason it might not be possible, based on what you know about the structure of the cell membrane. These "clues" will help you:
  — the protein portion of the membrane is thought to be involved;
  — proteins can move about within the membrane, although constraints imposed by the solubility rule apply;
  — proteins can change their shape, "exposing" previously hidden polar or nonpolar regions.

Answers:

Below, models of carrier-mediated transport proposed by scientists are described, with arguments for and against the feasibility of each. Model D is the one most widely accepted:

A. Diffusion of a membrane protein with a ligand through the membrane, shown in Figure 2–6.

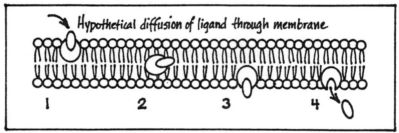

FIGURE 2–6

For: Proteins can diffuse through membranes. There is evidence that proteins of this type, called "ionophores," exist in certain bacteria and transport antibiotics across the membrane.

Against: A protein is made of both hydrophobic and hydrophilic amino acids. How does it shield itself from oppositely charged surfaces as it traverses the lipid bilayer? (It is possible that the protein protects itself by changing shape and shielding its hydrophobic and hydrophilic portions as needed.)

B. Rotation of a protein which spans the entire membrane as it carries the molecule being transported, as shown in Fig. 2–7.

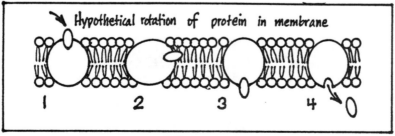

FIGURE 2–7

For: Some proteins do span the entire membrane.
Against: Same problem as in A.

C. "Bucket Brigade": Proteins arranged in tandem, spanning the width of the membrane, which transfer molecules to each other in turn by rotating or by passing the molecules through holes in themselves, as shown in Figure 2–8.

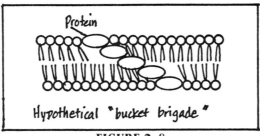

FIGURE 2–8

For: "Holes" in proteins do exist.
Against: Same problem as in A.

D. A polar channel exists within or between proteins which, accompanied by conformational change of the proteins, allows passages of molecules through the channel, as shown in Figure 2–9.

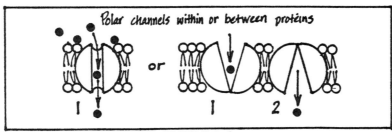

FIGURE 2–9

For: Some proteins do completely span the lipid bilayer. Hydrophilic channels within proteins do exist, allowing a hydrophilic molecule to be shielded from the hydrophobic bilayer. Proteins are known to undergo conformational changes.
Against: Scientists still do not understand the exact mechanism by which such a transfer could occur.

Evaluation

Method of Evaluation	Goal(s) tested
1. Question 1	1,2
2. Question 2	2
3. Question 3	3
4. Completion of student diagram	1,2,3
5. Participation in class discussion	1,2,3

Answer to Question 1:

Students' drawings should resemble those in the diagram below.

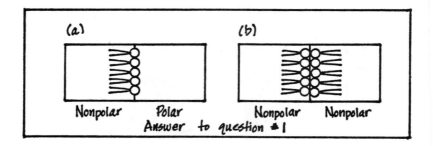

Answers to Question 2:

   1. NP   2. P   3. A

Answers to Question 3:

    1. 4   2. 2   3. 3   4. both

Name _____   Date _____

THE CELL MEMBRANE

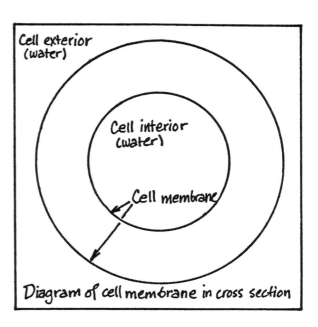

Diagram of cell membrane in cross section

Background

If a cell is viewed in cross section, it resembles the diagram at right.

Its membrane faces water on the exterior and on the interior. Molecules called *phospholipids* make up the basic structure of the cell membrane. How do these molecules arrange themselves so that the cell membrane is held together?

It turns out that phospholipid molecules contain two regions (pictured in the diagram below): the "head," which tends to orient towards water, and the "tail," which tends to move away from water.

Your Task

Follow the directions below to demonstrate how the phospholipid molecules arrange themselves so that all "water-loving" regions contact water, and all "water-hating" regions face away from water.

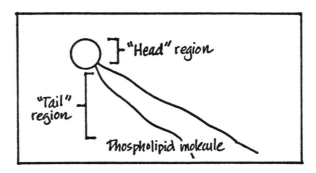

1. On the sheet labeled **CELL MEMBRANE IN CROSS SECTION**, label all regions which contain water.

2. Cut out about ten phospholipid molecules from the sheet provided by cutting along the straight lines.

3. Arrange the phospholipid molecules within the cell membrane cross section, making sure the "head" regions contact water and the "tail" regions face away from water.

4. Have your teacher approve your arrangement, then continue it by cutting out and gluing more phospholipid molecules around the cell membrane region.

CELL MEMBRANE IN CROSS SECTION

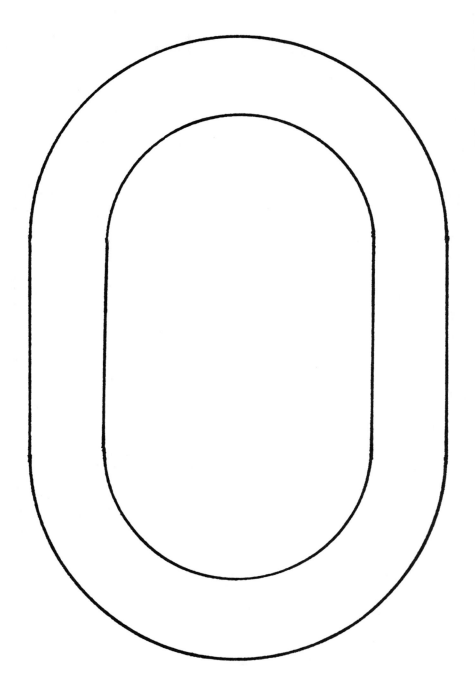

Teacher Approval ————————————————————

PHOSPHOLIPID MOLECULES

Cut along the straight lines.

THE CELL MEMBRANE EVALUATION QUESTIONS

Question 1

If phospholipid molecules were forced to separate the two substances in each box below along the line in the middle, how would they orient themselves so as to satisfy the "solubility rule"? Draw the proper arrangement along each line.

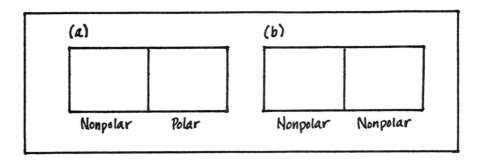

Question 2

You perform a laboratory experiment to determine whether various substances are polar, nonpolar, or amphipathic (having both polar and nonpolar character). Based on the results in the table below, determine the solubility of three unknowns in each of three solvents: water, oil, ethyl alcohol. Decide whether each unknown should be classified as polar, nonpolar, or amphipathic. Write the correct letters in the space given beside each unknown below using the following code:

P = polar   NP = nonpolar   A = amphipathic

UNKNOWN # ANSWER	WATER	OIL	ETHANOL
1.	insoluble	highly soluble	slightly soluble
2.	highly soluble	insoluble	slightly soluble
3.	slightly soluble	slightly soluble	highly soluble

THE CELL MEMBRANE EVALUATION QUESTIONS (continued)

Question 3

The diagram below represents a generalized diagram of the cell membrane in cross section. Answer the following questions about it by writing the appropriate number or word in the space provided:

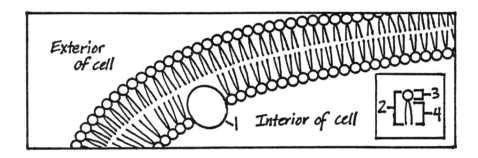

_____1. Which number refers to a portion of the membrane that is entirely nonpolar?

_____2. Which number represents an entire phospholipid molecule?

_____3. Which number refers to a portion of the membrane that is entirely polar?

_____4. Is the membrane protein drawn here entirely polar in character, entirely nonpolar, or both?

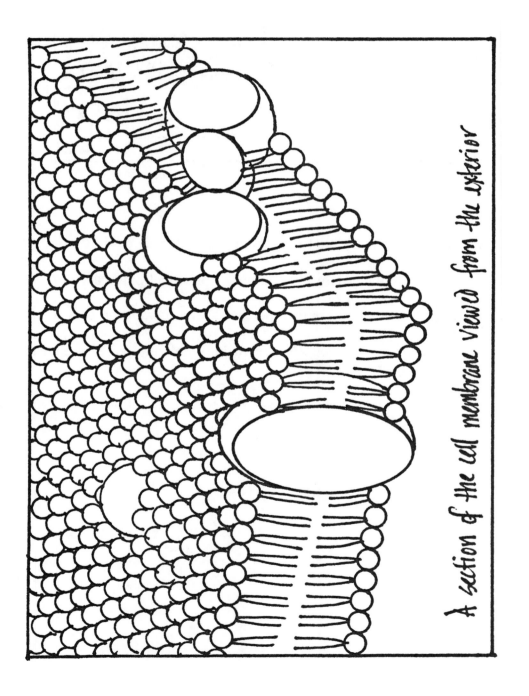

A section of the cell membrane viewed from the exterior

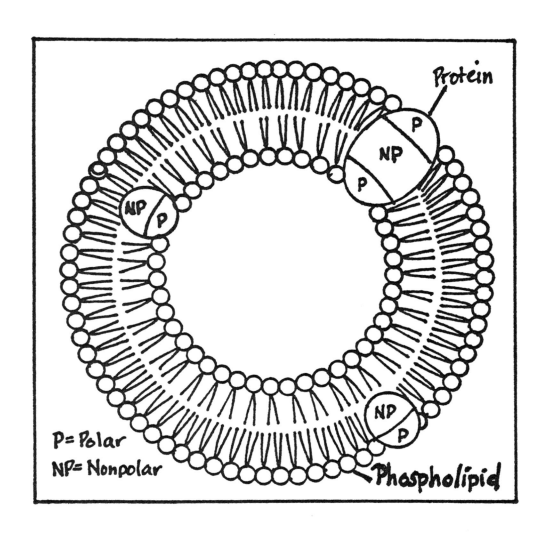

Protein

P

NP

P

NP

P

NP

P

P= Polar
NP= Nonpolar

Phospholipid

ACTIVITY 3 _____

WHY DOES A PLANT PIGMENT CHANGE COLOR?

Goals

After completing this activity, students will:

1. know the source of the colors of certain plant cells; and
2. be familiar with the effect of pH on the color of the plant pigment anthocyanin.

Synopsis

Students will form hypotheses about the effects of pH on the anthocyanin in purple cabbage leaves and test them by experiments they design.

Relevant Subjects

pH
cellular structure
plant reproduction and seed dispersal

Age/Ability Levels

grades 7–12, most ability levels

Entry Skills and Knowledge

Before participating in this lab, students should be able to:

1. define the terms *pH*, *acids*, and *bases*; and
2. pour and measure mild acids and bases safely in the laboratory.

Materials

Student worksheets: "Plant Colors," "Laboratory Planning and Data Sheet"

per pair of students
one glass marking pencil
purple cabbage (two large leaves)
distilled water (250 ml)
one beaker (500 ml)
cheesecloth (four layers, cut into 4 x 4″ squares) and funnel OR one kitchen tea strainer
five 20 ml test tubes
one graduated cylinder
five 5 ml pipettes
one pipette bulb

per class
blender
solutions (1 liter each) of different pH values
distilled water

24

DIRECTIONS FOR TEACHERS

Background on Anthocyanin

Anthocyanin is a pigment responsible for the blue, red, and purple colors of certain plants and is located in their vacuoles. It changes color as the surrounding pH varies. The pigment is purple in the vacuole of the purple cabbage cell, which has an internal pH of 6.8. Its colors at other pH values include the following:

1.0 – 3.5: red
3.5 – 6.0: rose   (these color changes are gradual)
6.0 – 7.0: purple
7.4 – 9.0: blue
9.0 – 13.0: green
13.0 – 14.0: yellow

SAFETY NOTE: Because extremely high and low pH solutions are dangerous for students, they should not be used in this lab. Nonetheless, great care should be exercised in handling ALL acids and bases. Flood with water if contact occurs with eyes or skin.

NOTE ON TERMINOLOGY: You will also find this cabbage referred to as "red" cabbage. To avoid confusion regarding the color at various pH values, however, the term "purple" is used here.

Preparation

Follow the steps below to make an extract of purple cabbage. It is your decision whether or not to let students make their own extracts.

1. Tear two large cabbage leaves into small pieces, discarding major veins.
2. Pour 250 ml of distilled water into a blender.
3. Add the cabbage to the blender and grind at medium speed for about 30 seconds.
4. Filter the extract through four layers of cheesecloth and funnel or a kitchen tea strainer into a beaker.
5. Discard what fails to pass through the cheesecloth.

Make up solutions (1 liter each) of different pH values using one of the suggestions below. Select pH values which will produce different colors in the cabbage extract. For example, make up solutions of pH values 5.7, 6.5, 7.4, then repeat colors within this range (such as 6.0, 6.8, 7.0, 7.2) so that students will have a variety of solutions to test.

- Use common "kitchen chemicals":
  vinegar (acetic acid, pH of 2.4–3.4)
  0.1 Normal solution of baking soda (sodium bicarbonate; a 0.1 N solution has a pH of 8.4).

(pH values of vinegar and baking soda © *Handbook of LaMotte Chemical Control Units for Science and Industry*, 13th ed., LaMotte Chemical Products Co., Chestertown, MD, 1944, used by permission.)

- Use commercially available buffer tablets.
- Make buffers according to the directions on pp. 160, 217, or 742 of Morholt and Brandwein (1986).

Photocopy student directions and laboratory sheets.

Teaching the Activity

1. Demonstrate to students the color change phenomenon by mixing 5 ml of purple cabbage extract solution with 5 ml of vinegar. The color will change from purple to red.

2. Indicate to students the contents of the tube containing the cabbage extract, but do not let them know yet what the other test tube contained. Ask for ideas as to why the solution changed color.

3. Accept all responses, then explain that vinegar was in the second test tube. Ask whether this information changes their ideas of why the phenomenon occurred.

4. If pH change is not offered as a response, tell students only that the pH of the solution had something to do with the color change.

5. Distribute student directions and lab sheets, briefly explain the assignment, and show students the available materials. It is your decision as to whether they may work individually or in pairs.

6. Provide assistance as needed, but ideas should come from students. Be sure all groups include a control with distilled water instead of a test solution. Three possible hypotheses and rationales are:

   - The purple cabbage extract will remain the color it became during the demonstration at all pH values because in my experience with acid-base indicators, I have never observed more than one color change in a substance.

   - The extract will retain its original purple color at all pH values other than the pH value used in the demonstration because it changes colors at this pH value; it is unlikely that it can change colors more than once.

   - The extract will change to colors other than purple and the color in the demonstration as the pH value is varied, because if it can change colors once, it should be able to change again. (This is the correct hypothesis.)

7. Check student lab preparation sheets, marking your approval on them before students begin their experiments.

8. When all students have completed the lab, assemble them as an entire group to share results. Make a chart on the board like the one on the data sheet for all groups to record their data.

9. Refer to the chart on the board as you and the students review the lab questions and those below. Answers to questions below are in parentheses.

   - What was the purpose of the control? (to provide a basis of comparison for color change, by controlling for the dilution of the color)

   - What is the pH value of the area of the cabbage cell in which anthocyanin is located? (6.0–7.0)

   - What would the pH values of the areas of these plant cells in which anthocyanin is located be?

purple onion (6.0–7.0)
blue cornflowers (8.0)
red geraniums (3.0)

- What is a function of plant pigments? What other pigments do plants have in addition to anthocyanin? (for attraction of animals for reproduction and seed dispersal, and for photosynthesis; chlorophyll, carotene, xanthophyll)

Answers to Questions on Student Worksheet

Sheet 1
I. Review

1. The degree of acidity or alkalinity of a solution (upper-level students should be able to tell you it is the negative log of the hydrogen ion concentration).
2. 14
3. a. basic
   b. acidic
   c. neutral
   d. basic
   e. acidic

Sheet 2
Answers will vary on questions 1–4; see teacher background for expected results at various pH values.

5. a. answers will vary
   b. anthocyanin
   c. yes, by comparing the color change produced by the unknown solution to the same color change produced by a solution of known pH value.

Follow-up Activities

- Have students find examples of plant use of pigments to attract insects and other animals for seed dispersal and reproduction, answering these questions:
  —Why do various insects respond to different colors?
  —What is the role of the animal in these plant processes?

- Have advanced students find out why the anthocyanin changed colors, answering these questions:
  —What is the shape of the anthocyanin molecule?
  —What is the molecular mechanism by which anthocyanin changes colors?
  —How do hydrogen ions interact with the molecule to cause this change? Why are there so many different color changes?

- Have students learn more about chemical indicators. Are their mechanisms of action similar to that of anthocyanin?

Evaluation

Method of Evaluation	Goal(s) tested
1. Completion of the student lab and data sheets	1,2
2. Participation in class discussion	1,2
3. Designing and testing an experiment, as described in this activity	1,2
4. Experiment 1	2

Experiment 1

Students are provided with materials used in the anthocyanin lab, but with the pH values of the test substances not labeled. They must use the materials they are given to determine the pH values of the unknown solutions and complete the following:

1. List the steps you will use to determine the pH values of the unknown solutions.

2. Record your results in the following table.

Number of Unknown Solution	pH Value

References

1. Sarquis, M., and Sarquis, J. *Institute for Chemical Education, Chem Is Fun: A Guidebook of Chemistry Activities for Elementary and Middle School Classes*, activity "The Cabbage Patch Detective;" University of Wisconsin, Madison (1989).

2. Morholt, E., and Brandwein, P. F., *A Sourcebook for the Biological Sciences* (3rd ed.). New York: Harcourt Brace Jovanovich, 1986.

3. Salisbury, F. B., & Ross, C. W., *Plant Physiology* (2nd ed.). Belmont, California: Wadsworth Publishing Company, Inc., 1978.

© 1991 by The Center for Applied Research in Education

Name _____ Date _____

PLANT COLORS

You have seen your teacher demonstrate what happens to the purple cabbage pigment anthocyanin in a solution at a particular pH value. Now you will design your own experiment to determine what happens to the color of this pigment at other pH values.

I. Review of pH

1. Define pH: _____

_____

_____

2. pH is measured on a scale from 1 to ___(fill in the blank).
3. Indicate whether each pH value below represents an acidic, neutral, or basic solution:

    a. pH 12.0 _____   d. pH 8.0 _____

    b. pH 2.0 _____   e. pH 4.0 _____

    c. pH 7.0 _____

II. Writing Your Hypothesis

Write a hypothesis on your *Laboratory Planning and Data Sheet* indicating what will happen to the color of the purple cabbage extract as the pH value of the solution is varied. Then, write a one sentence rationale for your hypothesis.

III. Designing Your Experiment

1. Your teacher will show you the materials that you may use to perform your experiment. Your only limitations are that:
   - you are allowed five test tubes per group, one of which must be a control, and
   - you must test solutions of four different pH values.

2. After reading the suggested procedure below, list on your *Laboratory Planning and Data Sheet* the steps you will follow in testing your hypothesis. Be sure also to indicate:
   - the contents of your control test tubes, and
   - the contents of your experimental test tubes.

3. Indicate in the table on your data sheet the anticipated final color of each tube.

4. Have your teacher approve your plan sheet, then perform your experiment and record the results in the table.

Suggested Basic Procedure

1. Label each test tube as to its contents.
2. Measure 5 ml of purple cabbage extract into each test tube.
3. One at a time, dispense 5 ml of the test solution into each test tube containing cabbage extract.
4. Record the color on your data sheet.

Name _____   Date _____

PLANT COLORS
LABORATORY PLANNING AND DATA SHEET

1. Hypothesis and rationale: _____

   _____

2. Experimental group(s): _____

   _____

3. Control group(s): _____

   _____

4. Procedure (List steps in the order you will perform them):

   _____

   _____

   _____

Tube Number	Contents	Anticipated Color	Actual Color
1			
2			
3			
4			
5			

5. Answer these questions:

   a. Was your hypothesis supported? If not, what hypothesis would you now test? Explain briefly your rationale for this hypothesis.

   _____

   b. What plant pigment is responsible for the purple color of cabbage leaves?

   _____

   c. Could you use purple cabbage extract to determine the pH value of a solution whose pH was unknown to you? Explain.

   _____

Teacher Approval _____

ACTIVITY 4

HOW WILL YOU DETERMINE THE EFFECT OF HEAT ON THE ACTIVITY OF AN ENZYME?

Goals

After completing this laboratory, students will:

1. understand the effects of temperature on enzyme activity; and
2. understand why fresh pineapple should not be used when making gelatin.

Synopsis

Students will form hypotheses regarding the effect of temperature on enzyme activity, then test them with experiments they design.

Relevant Subjects

biological molecules
enzyme action
nutrition

Age/Ability Levels

grades 9–12, most ability levels

Entry Skills and Knowledge

Before participating in this laboratory, students should be able to:

1. heat liquids safely in the lab; and
2. define the terms enzyme, substrate, active site, and product. It is also helpful, but not necessary, for students to be able to describe the lock-and-key and/or induced-fit models of enzyme-substrate action.)

Materials

student worksheets: "Determining the Effect of Heat on Enzyme Activity"

for the entire class:
powder funnel with filter paper OR tea strainer
timers or clocks
blender
fresh pineapple juice (unheated)
canned pineapple juice
refrigerator
model or diagram of enzyme-substrate action

per pair of students:
Six to nine test tubes (15 ml or 20 ml)
gelatin solution in 40°C water bath (50 ml)

One hot plate or Bunsen burner
ring stand and support if Bunsen burner is used
two or three 250 ml beakers
two pairs test tube clamps
two pairs safety goggles
three pipettes with bulbs (5 ml or 10 ml)
three droppers or Pasteur pipettes with bulbs
one graduated cylinder
one test tube rack
one marking pencil
one Celcius thermometer

DIRECTIONS FOR TEACHERS:

Background on the Pineapple Enzyme

Fresh pineapple contains an enzyme which degrades proteins. Since gelatin is a protein made from collagen, if fresh pineapple is used in making jello, the peptide bonds of the gelatin will be cleaved and the gelatin will not solidify. Canned pineapple can be used safely because it is heated prior to canning, rendering the enzyme inactive.

Preparation

1. Photocopy student worksheets.
2. Obtain laboratory materials.
3. Set up hot plates or Bunsen burners with beakers of water on them, but do not turn on the heat yet. Insert a thermometer into each beaker.
4. Prepare a demonstration by:
   a. filling several test tubes with 5 ml of gelatin alone;
   b. filling several other test tubes with 5 ml of gelatin and 5 drops of pineapple juice; and
   c. refrigerating the tubes overnight.
5. Prepare enzyme and substrate as follows:

   Enzyme:
   a. Peel and coarsely chop a third to a half of a fresh pineapple.
   b. Place pineapple chunks into a blender with about 1 cup of water (tap water can be used throughout this laboratory unless the water is highly acidic).
   c. Blend on high for about 30 seconds.
   d. Filter the pineapple mixture through a powder funnel and filter paper or a tea strainer. Discard pulp;
   e. Refrigerate pineapple juice until needed.

   Substrate:
   Prepare the substrate (gelatin) according to package directions.

Teaching the Laboratory

1. Review enzyme action with students using a commercial model or diagram as an aid.

2. Explain to students how the enzyme in fresh pineapple acts on gelatin.

3. Demonstrate the effect of fresh pineapple on gelatin by showing students the tubes you have prepared.

   As an alternative to steps 2 and 3 above, you may wish to encourage more active thought on the part of students by choosing this approach:

   a. Show students the prepared test tubes.
   b. Ask students for ideas to explain why some of the tubes solidified and others failed to.
   c. If students fail to suggest it, explain that there is an enzyme in pineapple which acts on gelatin, a protein. Ask them to figure out how this enzyme keeps the gelatin from solidifying.

4. Distribute lab sheets to students, show them the materials, and briefly explain their task. Have students work in pairs.

   SAFETY NOTE: Show students how to insert and remove test tubes from the beaker of water safely, using test tube clamps. Instruct them in how to turn on the Bunsen burner or hot plate correctly. Remind them to wear goggles at all times liquids are being heated and not to touch the beaker of water, thermometer, hot plate, or Bunsen burner and stand directly while they are hot.

5. As students design their experiments, be sure to:

   a. write your check of approval on the student data sheet when it is satisfactory;
   b. provide assistance as needed, especially regarding proper control groups (see below); most ideas, however, should come from students.

   Allow at least 20 minutes for designing experiments and at least 45 minutes for carrying them out.

6. Store labelled test tubes in the refrigerator overnight.

7. During the next class period, discuss results obtained by each group with the entire class by:

   a. making a data table on the chalkboard to include all of the different experimental and control groups;
   b. asking questions such as those below to help students draw a conclusion about the effect of increasing temperature on enzyme activity:

   ● Why did gelatinization occur after heating the fresh juice at high temperatures, but not when heated at lower temperatures?

   ● Why was there partial gelatinization at intermediate temperatures?

   (At this point you might ask students to construct graphs correlating reaction rate to temperature and degree of gelatinization to temperature.)

   The *correct explanation* is as follows:

   Heat increases the rate of an enzyme-substrate reaction up to a point. This is due to an increase in molecular motion which causes collisions between enzyme and substrate molecules to occur more frequently. After this point (around 50°C for most enzymes), the reaction ceases because the enzyme is rendered inactive by denaturation. Because of this, there will probably be some tubes heated to a temperature slightly below 50°C which only partially jell, indicating partial enzyme destruction. A correct graph of enzyme reaction rate versus temperature is shown in Figure 4–1.

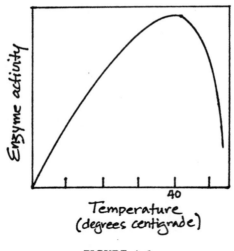

FIGURE 4-1

    c. Reviewing answers to questions on student lab sheets.
8. Collect student lab sheets to grade.

Control groups which should be run include:

1. 5 ml of gelatin alone
2. 5 ml of gelatin plus 5 drops of water (to control for the effects of dilution)
3. gelatin with 5 ml untreated fresh pineapple juice (if fresh juice is used in the experiment)
4. gelatin with 5 ml untreated canned pineapple juice (if canned pineapple juice is used in the experiment)

Sample student hypothesis, rationale, and data table

Hypothesis: If fresh pineapple juice is heated at 100°C for 15 minutes, mixed with gelatin and stored at 4°C for 24 hours, then gelatinization will not occur.

Rationale: Since gelatin is heated in order to prepare it, heating should have no effect on whether it jells. Heating the pineapple juice should not affect its activity on gelatin, since enzymes should not be destroyed this easily.

Answers to student questions 1–5 will vary.

TABLE 4–1

TREATMENT	SAMPLE NUMBER	RESULTS

Experimental groups:
I. gelatin + fresh
   pineapple juice
   (heated)
	1
	2
	3
		+
		+
		+

Control groups:
I. gelatin + fresh
   pineapple juice
   (untreated)
	1
	2
	3
		–
		–
		–

II. gelatin alone	1 2 3	+ + +
III. gelatin + water	1 2 3	+ + +

Follow-up Activities

- A limitation of this experiment is that it is impossible to witness the enzyme-substrate reaction as it occurs; only the results can be assessed. Ask students how a related enzyme experiment could be set up so that the product could be measured during the reaction.

- Obtain a commercial or student-made movable model (made of wire, for example) of an enzyme, showing its active site and interaction with its substrate. By twisting and unraveling it, simulate what happens when an enzyme is denatured. It should be evident why denaturation prevents the enzyme from reacting with its substrate.

- Have students perform the experiment described in this activity with fresh frozen pineapple juice.

- Have students modify the gelatin pineapple activity to determine the effect of pH on enzyme activity by mixing the following with the gelatin, the enzyme, or both. They will have to experiment to determine effective concentrations:

  —vinegar or lemon juice

  —baking soda solution

  —buffers at different pH values

- Have students observe the effects of heat and low pH on protein in this way:

  Pour an egg white into boiling water, and another into vinegar at room temperature. Explain why the contents of the two beakers appear the same.

- Have students perform the gelatin-pineapple experiment varying the amount of time the gelatin is heated, but keeping the temperature constant for all experimental groups.

Evaluation

Method of evaluation	Goals tested
1. Ask students to interpret their data (see Table 4–1).	1,2
2. Completion of student lab sheets.	1,2
3. Designing and performing an experiment as described in this activity.	1,2
4. Participation in class discussion.	1,2

DETERMINING THE EFFECT OF HEAT ON ENZYME ACTIVITY

You are to design an experiment to determine the effect of heat on the activity of the enzyme in fresh pineapple. Your teacher will provide you with background information about the enzyme and its effect on gelatin and show you what materials you may use to perform this lab.

Writing a Hypothesis and Designing the Experiment

1. Answer the following questions about your experimental procedure, then include these answers in your hypothesis.

    In your experiment:

    a. Do you plan to use fresh pineapple juice, canned, or both?

    _____

    b. Will you change the temperature of the pineapple juice, the gelatin, or both? What will this temperature be? (You should keep your experiment at the chosen temperature for at least 15 minutes.)

    _____

    c. How will you determine the results after temperature treatment?

    _____

    A sample hypothesis might be:
       If canned pineapple juice is heated to 100° for 15 minutes, then mixed with gelatin
       and stored at 4°C for 24 hours, then gelatinization will not occur.

2. Write your hypothesis and a one-sentence rationale for it.

    _____

    _____

    _____

    _____

3. Read the laboratory procedure and limitations below, then list the steps you will follow as you test your hypothesis. Be sure to identify:
    a. the variable(s)
    b. the experimental group(s)
    c. the control group(s)

Laboratory Procedure

For each test tube you work with,
   a. Measure 5 ml of gelatin;
   b. Add 5 drops of pineapple juice;
   c. Swirl to mix;
   d. Label the contents.

In order to heat tubes, place them in a beaker of water on a hot plate or Bunsen burner as directed by your teacher. *SAFETY NOTE:* **Wear goggles at all times tubes are being heated. Handle tubes only with test tube clamps. Do not touch the beaker, thermometer, hot plate, or Bunsen burner and stand directly while they are hot.**

DETERMINING THE EFFECT OF HEAT ON ENZYME ACTIVITY (continued)

Limitations:

a. You must include at least one experimental and one control group;
b. Triplicates (at least) of each experimental and control group should be made (several groups may share control samples); and
c. You are allowed a maximum of nine test tubes per group.

Your Experiment:

_____

_____

_____

Setting Up a Data Table

Set up and complete a data table, to be handed in for a grade, similar to Table 4–2. You will record the data in your table. After your teacher approves your plan, perform your experiment. Afterwards, answer the questions below.

Questions:

1. What results did you expect? _____

_____

2. Did your results support or contradict your hypothesis? Explain.

_____

_____

3. What was the purpose of including each of your control groups?

_____

_____

4. What were the possible sources of experimental error?

_____

_____

5. Based on class data:
   a. What is the effect of increasing temperature on the activity of this enzyme?

   _____

   _____

   b. How do you think the canned pineapple was treated prior to canning? What information leads you to this conclusion?

   _____

   _____

TABLE 4–2

TREATMENT	SAMPLE NUMBER	RESULTS
Experimental groups: I. (briefly describe procedure)	1 2 3	Indicate with a + or − in the proper column whether or not gelatinization occurred
II.	1 2 3	
Control groups: I.	1 2 3	
II.	1 2 3	
III.	1 2 3	

section II

Plant Structure and Function

Many students seem to hold the view that plants might as well be inanimate objects. Since they don't "do" anything, I have heard students comment, plants must be very boring to study.

On the contrary—plants are fascinating! The two activities included here illustrate processes and structures unique to plants, demonstrating not only that plants are active physiologically, but also that there are dynamic interactions between plants and the environment.

Both activities in this section require students to deduce principles for themselves, either through observation alone (in "How Are Seeds Dispersed?"), or through designing and testing experiments (in "What Are Some Factors Affecting Transpiration?").

The seeds dispersal activity has students determine solely by examining the structure of the seed some interactions between the plant and its environment which assure that its offspring will be dispersed, increasing the chances of survival of the species.

"What Are Some Factors Affecting Transpiration?" requires that students design and test an experiment as they observe a plant actively releasing water to the environment.

Other activities in this book which make use of interesting phenomena involving plants include the following:

1. "How Will You Solve the Plant Cell's Problems?"
3. "Why Does a Plant Pigment Change Color?"
4. "How Will You Determine the Effect of Heat on the Activity of an Enzyme?"
12. "How Are Traits Transmitted in Corn Plants?"
13. "What Are Some Interactions Between Genes and Environment?"
14. "What Is the Effect of Radiation on Plants?"

How Should These Activities Be Used?

Some suggestions for integrating these activities into your biology course include the following:

1. "How Are Seeds Dispersed?" can follow a lesson on plant life cycles. It may also be used to introduce a unit on interactions between organisms and the environment.
2. "What Are Some Factors Affecting Transpiration?" might follow a discussion of photosynthesis which includes the importance of gas exchange with the environment. Since student-designed experiments will reveal environmental adaptations affecting the rate of transpiration, the activity is a way of introducing this topic.

ACTIVITY 5

HOW ARE SEEDS DISPERSED?

Goals

After completing this activity, students will:

1. be familiar with mechanisms of seed dispersal;
2. know taxonomic names of some common plants; and
3. better understand how certain plants have adapted to the environment.

Synopsis

Students will observe some common seeds and discover their mechanisms of dispersal.

Relevant Topics

plant life cycles
plant structure and function
taxonomy
environmental adaptation

Age/Ability Levels

grades 7–11, most ability levels

Entry Skills and Knowledge

Before participating in this activity, students should be able to:

1. explain the importance of the seed to the plant;
2. state the plant structure that develops into a seed;
3. list common methods of seed dispersal; and
4. distinguish between scientific names and common names of living things.

Materials

student worksheets: "Seed Dispersal Lab Directions and Review," "Seed Dispersal Laboratory
    Sheet"
sections of cardboard (approximately 17 inches by 22 inches)
felt-tip markers
various types of seeds
tape
plant classification manuals

DIRECTIONS FOR TEACHERS

Preparation

1. Photocopy student direction and laboratory sheets. You will need to duplicate two or
   three copies of the lab sheet for each student.
2. Obtain materials for stations, including seeds. Readily available seeds are listed below,
   with their mechanisms of dispersal in parentheses. You may wish to assign students
   to bring in seeds.

—ash (wind)
—elm (wind)
—acorn (edible, gravity)
—maple (wind)
—magnolia (edible, wind)
—sycamore (gravity, wind)
—dandelion (wind)
—coconut (gravity, water)
—burr (hitchhiker)
—catalpa (mechanical forces)
—peach pit (edible, gravity)
—pine (wind, gravity, mechanical forces)
—bean (mechanical forces)
—pea (mechanical forces)
—blackberry (edible)
—beggar-tick (hitchhiker)

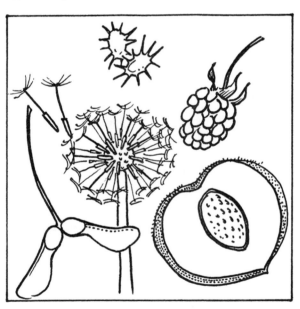

3. For each seed, print in felt-tip marker on a piece of cardboard:
   —a station number;
   —its common name; and
   —its scientific name (use classification manuals if you need help finding these).

4. Secure one cardboard piece at each "station"—a flat surface such as a lab bench—using tape. Fasten one seed to each piece of cardboard; have other seeds free for student examination. Create enough stations so that two students can work at each simultaneously.

5. Be sure students have mastered the entry skills.

SAFETY NOTES:

● If you use any seeds which are treated with pesticides or fungicides, wash them before using.

- Caution students not to eat seeds and to keep all plant parts out of their mouths. Have them wash their hands at the end of each lab period.
- Check your local or state safety manual for a list of plants with toxic seeds; these should not be used in this lab. In particular, avoid plants bearing red or white fruits.

Teaching the Laboratory

1. Distribute student directions and lab sheets, and briefly explain the purpose of the lab. Ask questions such as those below to help students begin thinking about seed dispersal. Answers are in parentheses.
   - What would happen if all the seeds from an adult plant, instead of being dispersed, attempted to grow near the parent plant? (There would be competition for available light and nutrients, and many plants would not survive.)
   - If a seed were dispersed by "hitching a ride" on a dog's fur, what might the appearance of the seed be? (It might have projections allowing it to attach to the fur.)
   - What would be the probable weight of a seed dispersed by wind, light or heavy? (light); one dispersed by gravity? (heavy)
2. Have students complete the review questions on their direction sheet prior to the lab, either at home or in class. It is your decision whether to allow them to use their books in answering the questions.
3. Have students work in groups of two as they travel from station to station, completing the information on their lab sheets. Allow at least three minutes at each station, and have groups change stations simultaneously.
4. When all students have completed their laboratory sheets, perhaps on the following day, hold a class discussion of their work. Ask questions such as those below to stimulate thought:
   - Describe some of the ways you think these seeds are dispersed, and explain what characteristics of the seeds led you to those conclusions. (Answers will vary.)
   - What are some common structural features among all the seeds dispersed by wind? (Answers will vary; repeat this question for other methods of dispersal.)
   - Why could the coconut not be dispersed easily by wind alone? (It is heavier than many other types of seeds.)
   - Explain how the seeds you observed have adapted to the environment. (Answers will vary.)
5. Collect and grade laboratory sheets.

Answers to Review Questions

1. ovule
2. adult plant
3. they represent embryo plants which continue the species
4. Answers may come from the following list:
   a. wind
   b. water
   c. hitchhiking
   d. being encased in an edible fruit

e. gravity

f. mechanical forces

Follow-up Activities

- Set up stations as described in this activity with fungi instead of seeds. Have students determine through observation how the fungi reproduce. Ask them also to:

  —point out on their drawings where the spores are located on each fungus; and

  —explain how the spores are dispersed.

- Have students "create" an environment on another planet, then "invent a seed dispersed by _____" on that planet. Have them fill in the blank, and make drawings of seeds in the "foreign" habitats. Afterwards, they can explain to the class how seeds are adapted to the enviornment.

- Have students view prepared slides under the microscope of plant structures such as the following:

  —seeds in cross section

  —pollen grains

  —ovules in cross section

  —sections of mature ovaries in cross section

- Have students bring in fruits from home and dissect them, identifying on drawings the area of each fruit which:

  —developed from the ovary

  —developed from the ovule

  —can become a mature plant

  Afterwards, enjoy eating the fruits!

SAFETY NOTES:

- Do not allow students to eat fruits in the laboratory or classroom where chemicals, pathogens, and mold spores have been used.

- Make sure students "dissect" fruits to be eaten only with knives which have not been used in laboratory procedures.

Evaluation

Method of Evaluation	Goal(s) Tested
1. Completion of Student lab sheets	1,2,3
2. Participation in class discussion	1,3

References

1. Maryland State Department of Education, *Maryland School Science Safety Project*: *Safety Manual K–12*, 1988.

2. Towle, A. *Modern Biology*. New York, N.Y.: Holt, Rinehart, and Winston, 1989.

SEED DISPERSAL LAB DIRECTIONS AND REVIEW

You will examine some common seeds and attempt to discover through observation how they are dispersed.

Answer the review questions below, then travel from station to station, as instructed by your teacher. As you do, complete the information on the Seed Dispersal Laboratory Sheet.

SAFETY NOTES:

● Keep all plant parts away from your mouth. Do not eat seeds; some may be treated with toxic pesticides or fungicides.

● Wash your hands after each lab in which you handle seeds.

Review

1. From what plant structure does the seed develop?

_____

2. What structure will the seed become upon germination?

_____

3. Why must seeds be dispersed?

_____

4. List and describe briefly three methods of seed dispersal:

a. _____

b. _____

c. _____

Name _____ Date _____

SEED DISPERSAL LABORATORY SHEET

Station #: _____
Common Name: _____
Scientific Name: _____
Probable Means
of Dispersal: _____
Diagram:

Station #: _____
Common Name: _____
Scientific Name: _____
Probable Means
of Dispersal: _____
Diagram:

Station #: _____
Common Name: _____
Scientific Name: _____
Probable Means
of Dispersal: _____
Diagram:

Station #: _____
Common Name: _____
Scientific Name: _____
Probable Means
of Dispersal: _____
Diagram:

Station #: _____
Common Name: _____
Scientific Name: _____
Probable Means
of Dispersal: _____
Diagram:

Station #: _____
Common Name: _____
Scientific Name: _____
Probable Means
of Dispersal: _____
Diagram:

ACTIVITY 6

WHAT ARE SOME FACTORS AFFECTING TRANSPIRATION?

Goals

After completing this activity, students will:

1. better understand the process of transpiration in plants;
2. understand some factors which affect transpiration; and
3. be familiar with some mechanisms of adaptation by plants to the environment.

Synopsis

Students will form hypotheses to determine factors affecting transpiration in various plants, then design experiments to test them. The color change in cobalt chloride paper will be used as an indicator of transpiration.

Relevant Topics

environmental adaptation
photosynthesis
plant anatomy and physiology

Age/Ability Levels

grades 7–12, most ability levels

Entry Skills and Knowledge

Before participating in this activity, students should be able to:

1. describe briefly the process of transpiration; and
2. name the plant structure through which transpiration occurs.

Materials

student worksheets: "Transpiration Laboratory Preparation," "Transpiration Laboratory Data," and "Plant Transpiration Evaluation Questions" (optional)

per pair of students
various types of potted and outdoor plants
one pair scissors
twelve paper clips
twelve small squares plastic wrap or plastic bags
twelve squares plastic wrap or plastic bags
twelve squares cobalt chloride paper
six rubber bands

DIRECTIONS FOR TEACHERS

Background

- Cobalt chloride paper is blue when dry and pink when moist. This color change can therefore be used as a measure of the presence and degree of transpiration.

49

- For transpiration to occur and produce a color change, the following must be true:
  —there must be stomata on the area of the plant being tested; and
  —the stomata must be open during the experiment.
- Students will be monitoring:
  —whether there is a color change;
  —the amount of time required for the paper to change color; and
  —the richness of the pink color.
- Factors affecting transpiration include these:
  —The species of the plant tested determines the number of stomata present, their arrangement, and their location on the plant. For example, variations in numbers and arrangement of stomata on the surfaces of leaves are listed below:
  —In general, the undersides of leaves have many more stomata than the upper surfaces. Examples of these kinds of leaves are: pumpkin, tomato, bean, poplar.
  —Some leaves have no stomata on the upper surface, such as: balsam fir, red Begonia, rubber plant, lily, lilac, nasturtium.
  —Leaves with almost equal numbers of stomata on both surfaces include: oats, sunflower, corn, cabbage.
  —Floating plants such as the water lily have no stomata on the lower surface.
  —Leaves tend to have more stomata than stems.
  —Some plants have fewer stomata on their "shade" leaves than on their "sun" leaves.
  —Arid plants (xerophytes) have fewer stomata than nonarid plants (mesophytes).
  —CAM (Crassulacean acid metabolism) plants such as cacti, *Kalanchoë*, and *Sedum* differ from other plants in that their stomata are open at night and closed during the day.
  —Some environmental variables affect whether stomata are open or closed. Each of the following three factors increases the rate of water loss of a plant, which in turn decreases turgor in the guard cells, closing the stomata: high air temperature, great amount of sunlight (by raising air temperature), low level of humidity.
  —Factors that decrease the amount of water entering the plant, also causing the stomata to close, include: low level of soil moisture, and high level of soil salinity.
  —Likewise, these factors cause an opening of the stomata: low air temperature, low level of sunlight, high level of humidity, high level of soil moisture.

Preparation

1. Obtain commercially prepared cobalt chloride paper, and cut into small pieces, 2 inches by 2 inches. Dry them briefly in a warm oven if they are pink.

   Or: Make your own cobalt chloride paper by preparing a 5% solution of cobalt chloride in distilled water (5 g cobalt chloride in 100 ml total volume) dipping strips of filter paper into the solution, and air drying them on paper towels.

2. Obtain potted plants representing as many of those species described in the background section as possible. Several students may share one plant. Consider using outdoor plants, as well.

Teaching the Lab

1. Briefly explain the purpose of the lab, then distribute student directions and data sheets.

   SAFETY NOTES:
   - Students should not ingest cobalt chloride.

- Students should never place any plant in their mouths; have students wash hands at the end of each lab period when plants are used.
- Check your state or county safety manual for a list of plants which are toxic; these should not be used in this lab. Examples of potentially toxic plants are hydrangea and tomato. One rule of thumb is to avoid using plants which bear red or white fruits or which have a milky, white juice.

2. Provide students with a list of the plants they can use. You may need to describe a sample hypothesis and experimental design for them, such as the following:

- **Hypothesis:** The upper surfaces of the leaves of a nondesert plant will transpire more than their lower surfaces.

   Experimentals: Three cobalt chloride strips on the upper surfaces of leaves of the plant.

   Controls: Three strips on the lower surfaces of the plant.

- **Hypothesis:** The leaves of a desert plant will transpire more than will the stems.

   Experimentals: Three strips of cobalt chloride paper on the upper surface of each leaf; three on the lower surface.

   Controls: Three strips on the lower leaf surfaces, measured at night.

- **Hypothesis:** The lower surface of a leaf transpires more at midday than at night.

   Experimentals: Three cobalt chloride strips on the lower leaf surfaces, measured at midday.

   Controls: Three strips on the upper leaf surfaces, measured at night.

   NOTE: For this experiment, measurement must be done quickly, and with as little light as possible.

- **Hypothesis:** The leaves of a tree which are exposed directly to the sun transpire less than those in the shade.

   Experimentals: Three strips of cobalt chloride paper on the upper surface of each "sun" leaf; three on the lower surface.

   Controls: Three strips on the upper surface of each "shade" leaf; three on the lower surface.

- **Hypothesis:** The leaves of a desert plant transpire less than those of a nondesert plant.

   Experimentals: Three strips of cobalt chloride paper on the upper leaf surface of each desert plant; three on the lower surface.

   Controls: Three strips on the upper leaf surface of each nondesert plant; three on the lower surface.

3. Have students work in groups of two as they design their experiments.
4. Mark your "OK" on student lab sheets, making sure that:
   - Each experimental and control group is planned in triplicate;
   - A proper control group is included;
   - Students have anticipated and solved problems, such as a means to experiment on plants at night; and
   - The same types of plants are used in both experimental and control groups.
5. As students collect their data, have them record results on a chart on the chalkboard.

6. Compare results as a class. The following questions may stimulate discussion. Answers are given in parentheses.

   Why is it that:

   • water plants have stomata only on the upper leaf surfaces? (gas exchange would be blocked by water if they were on the lower surface)

   • arid plants have fewer stomata than nonarid ones? (to conserve water in an environment with high temperatures and low humidity)

   • "shade" leaves have fewer stomata than "sun" leaves? (the leaves in direct contact with the sun photosynthesize more and are in greater need of a means of gas exchange)

   • some desert CAM plants open their stomata only at night, as compared with the daylight opening of most plants? (to conserve water in the higher temperatures and lower humidity of the daylight hours)

Answers to Questions on Student Lab Sheets

Introduction

1. pink

Preparation

1. a. stomata

   b. leaves, stems

   c. see teacher background section

Follow-up Activities

• Show students a commercial model of a leaf in cross section. Ask them what relationships the spongy layer and stomata have.

• Have advanced students research Crassulacean Acid Metabolism to learn more about the metabolic advantage to these plants of opening their stomata only at night. It involves an enzyme, ribulose-bis-phosphate carboxylase, which is inactive in the dark.

• Have students observe prepared slides of stomata from different plants.

• Have students compare micrographs of stomata from desert plants with those of non-desert plants. To prevent water loss, desert stomata are usually sunken, the surrounding tissue has thick cutin, and they are covered by projections of other cells.

• Have students observe stomata in cells of some succulents, such as *Kalanchoë*, *Sedum*, and *Peperomia*, by gently tearing away a layer of cells of the lower surface of a broken leaf. The stomata may be observed at different concentrations of NaCl to determine how turgor affects them.

Evaluation

Method of Evaluation	Goal(s) Tested
1. Designing and performing an experiment, as described in this activity	1,2,3
2. Participation in class discussion	1,2,3
3. Completion of student worksheets	1,2,3
4. Evaluation Questions A (1–8) and B	1,2,3

Evaluation question A answers: (See teacher's section for explanations.)

1. b	5. a
2. a	6. b
3. b	7. b
4. b	8. a

Question B answers   a. N   b. A   c. N   d. A

References

1. Galston, A. W., Davies, P. J., Satter, R. L., *The Life of the Green Plant* (3rd ed.). Englewood Cliffs, N.J.: Prentice-Hall, Inc., 1980.

2. Maryland State Department of Education, *Maryland School Science Safety Project: Safety Manual K-12,* 1988.

3. Morholt, E., Brandwein, P. F., *A Sourcebook for the Biological Sciences* (3rd ed.). New York, N.Y.: Harcourt, Brace, Jovanovich, 1986.

4. Duggar, B. *Plant Physiology,* New York, N.Y.: Macmillan, 1930.

Name _____ Date _____

TRANSPIRATION LABORATORY PREPARATION

You are to form a hypothesis about a factor affecting transpiration in plants and design an experiment to test it. Evidence of transpiration can be seen by attaching cobalt chloride paper to plants and watching for a color change. The paper is blue when dry and pink when moist.

1. What color will the paper be if transpiration is occurring? _____

Preparation

1. Answer these questions:

   a. Through what structures on the plant does transpiration occur?

   _____

   b. On what area(s) of the plant are these structures located?

   _____

   c. The amount of water leaving a plant is regulated by the guard cells of the stomata. Choose THREE of the variables below and explain in Table 6–1 how each might affect the regulation of the rate of transpiration.

   variables:

   - type of plant (a desert plant vs a non-desert one)
   - location of the stomata on the plant (stems vs leaves; upper vs lower surfaces of leaves)
   - air temperature
   - time of day
   - level of soil moisture
   - level of soil salinity
   - humidity
   - amount of available sunlight

TABLE 6–1

	VARIABLE	ROLE IN REGULATION
1.		
2.		
3.		

Writing a Hypothesis and Designing the Experiment

1. Choose ONE of the variables from Table 6–1 around which to design your experiment. On your Laboratory Data Sheet, indicate which one you have chosen.
2. Write your hypothesis on your lab sheet, and write a one-sentence rationale for it.
3. Read the following Suggested Basic Procedure, then list the steps you will follow in testing your hypothesis under "Your Experimental Procedure." Your teacher will provide you with a list of plants with which to work. In Table 6–2, indicate:

   - your experimental group(s)
   - your control group(s)
   - expected results

Your Experimental Procedure:

_____

_____

_____

_____

_____

4. Obtain teacher approval before beginning the experiment. Afterwards, answer the questions on your data sheet.

Suggested Basic Procedure

1. Cut cobalt chloride paper into smaller strips, if necessary.
2. Attach small pieces of cobalt chloride paper to each of the plants you are going to test. Cover each with a piece of plastic and secure it carefully with a paper clip (if on a leaf) or with a rubber band (if on a stem or very thick leaf). This is shown in Fig. 6–1.

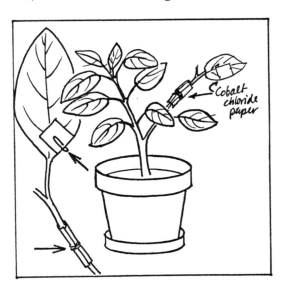

Cobalt chloride paper

FIGURE 6–1. ATTACHMENT OF COBALT CHLORIDE PAPER.

SAFETY NOTES:

- Never ingest cobalt chloride.
- Keep all plants away from your mouth, as some are toxic. Always wash hands after each lab involving plants.

3. Watch the paper for no longer than 30 minutes. Record in Table 6–2 the amount of time required for a color change and the final color.

Name _____ Date _____

TRANSPIRATION LABORATORY DATA SHEET

1. Variable: _____
2. Hypothesis: _____

_____

3. Rationale: _____
_____

Teacher Approval _____

TABLE 6–2

Group		Expected Results	Actual Results Color	Time
Experimentals:				
#1	1			
	2			
	3			
#2	1			
	2			
	3			
Controls:				
#1	1			
	2			
	3			
#2	1			
	2			
	3			

3. Questions

   a. Was your hypothesis supported? If not, what hypothesis would you now test? Briefly explain your rationale for this hypothesis.

   _____

   _____

   b. How do you explain the results of your experiments? _____

   _____

   _____

PLANT TRANSPIRATION
EVALUATION QUESTIONS

A. In which situation below is more transpiration likely to occur, in *a* or in *b*? Circle the correct response, then explain why.

    1. a. a cactus during daylight
       b. a cactus at night
       Reason:

    2. a. the outer leaves of a maple tree
       b. the inner leaves of a maple tree
       Reason:

    3. a. a *Kalanchoë*
       b. a geranium
       Reason:

    4. a. the stem of a nasturtium
       b. the leaves of a nasturtium
       Reason:

    5. a. the upper surface of a water lily leaf
       b. the lower surface of a water lily leaf
       Reason:

    6. a. the upper surface of a rubber plant leaf
       b. the lower surface of a rubber plant leaf
       Reason:

    7. a. the upper surface of a tomato leaf
       b. the lower surface of a tomato leaf
       Reason:

    8. a. the upper surface of a sunflower leaf
       b. the upper surface of a balsam fir leaf
       Reason:

B. Plants from earth are being sent to populate another planet, one known to have an environment similar to the desert of the American southwest. Select the characteristics that would give these plants the best chance of surviving in their new environment. Use the following code:

    A—characteristic that will enable adaptation to new environment
    N—characteristic that will NOT enable adaptation to new environment

    ————————a. The plant has no roots.
    ————————b. The leaves of the plant have a waxy substance on the exterior.
    ————————c. There are many stomata on the leaves of the plant.
    ————————d. The stomata are able to open only at night.

section 3

Organisms Interacting with the Environment

The activities in this section are designed to help students understand some dynamic interactions occurring among organisms and the environment. The first three activities deal with how the presence or absence of one organism in an ecosystem can affect not only other living creatures within that environment, but the nonliving components as well. Likewise, the effects of the nonliving aspects of an environment on its living organisms are considered.

The last two activities require students to consider specific adaptations of organisms which enable them to survive in their particular environment.

All activities stress the process of science in various ways. "Is the Water Around You Safe to Drink?" and "What Are the Effects of Human Intervention on a Food Web?" require students to make predictions regarding biological interactions. In the first, students test their predictions with experiments they design and carry out themselves, as they analyze the effect of human beings and other organisms on various sources of water. In the latter, students compare their proposed solutions to a problem with that of scientists.

"What Is a Food Pyramid?" and "How Is an Organism Adapted to Its Environment?" require students to discover principles and relationships existing in nature. The first accomplishes this goal by having students create a simulated food pyramid. The second has students observe organisms in their natural habitats.

Finally, "Can You Design a New Life Form?" has students apply principles of interaction within an environment as they "create" a hypothetical planet, and design a life form suited to live in that environment.

How Should These Activities Be Used?

Below are some suggestions for integrating these activities into a traditional high school biology course.

1. Incorporate into a unit on ecology "What Is a Food Pyramid?" and "What Are the Effects of Human Intervention on a Food Web?."

2. "Is the Water Around You Safe to Drink?" can serve as a laboratory in a unit in either microbiology or ecology. Interest in this lab is usually quite high, since students can test familiar sources of water for contamination, such as water faucets at home and at school.

3. "How Is an Organism Adapted to Its Environment?" and "Can You Design a New Life Form?" can both be integrated into a unit on evolution. Make certain that students have observed examples of adaptations of organisms to their environments on earth before they are asked to design their own life forms. They can observe these adaptations either by carrying out "How Is an Organism Adapted to Its Environment?," or by watching films of unusual adaptations.

ACTIVITY 7

WHAT IS A FOOD PYRAMID?

Goals

After completing this activity, students will:

1. understand the concept of a food pyramid;
2. understand more about interactions within a population; and
3. understand why "eating lower on the food chain" enables an organism to have more food available.

Synopsis

Students will discover the shape of a food pyramid by "constructing it" on paper using a hypothetical food chain.

Relevant Topics

trophic levels
population growth
food chains

Age/Ability Levels

grades 7–10, most ability levels

Entry Skills and Knowledge

Before participating in this activity, students should be able to:

1. define the term *food chain;*
2. define the terms *producers* and *primary*, *secondary*, and *tertiary consumers*;
3. use a metric ruler correctly; and
4. describe predator-prey relationships;

Materials

student worksheet: "What Is a Food Pyramid?"
metric rulers
optional: "Food Pyramid Evaluation Questions" worksheet

DIRECTIONS FOR TEACHERS

Preparation

1. Obtain metric rulers.
2. Photocopy student worksheets.

Teaching the Activity

1. Be sure students have mastered entry skills.
2. Briefly explain the assignment to students. Distribute student worksheets and rulers.

3. It is your choice whether to have students work individually or in groups of two. Provide time for them to complete their worksheets, answering questions as needed.

4. When all students have completed the assignment, call for a volunteer to draw the pyramid shape on the board and review the answers to questions on the worksheet.

Answers to Student Questions

1 and 2. See the sample diagram shown in Fig. 7–1.

3. pyramid, or triangle

4. a. vorteks, because they convert solar energy into biochemical energy
   b. snives, because they consume the producers
   c. klukes, because they prey upon organisms which consume producers

5. Energy is lost in respiration from one trophic level to the next, so fewer numbers of organisms exist on each level.

6. a large tree which supports many insect primary consumers, with many parasites on the insects

7. a. All three would disappear.
   b. The number of snives would increase because they would have no predators. The number of vorteks would then decrease because more would be consumed. The snives would die as they depleted their food supply; the number of vorteks would then increase, and the cycle would repeat.

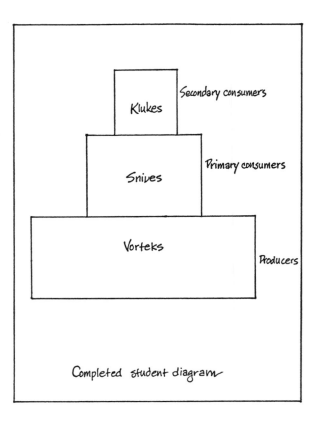

c. The number of vorteks would decrease because more would be consumed; the number of snives and klukes would decrease due to a lack of food. The skunkola population, however, would increase.
   d. There would be an increase in the number of vorteks because of a lack of predators. The number of klukes would decrease due to a lack of food.
   e. The number of snives and klukes would increase.
   f. The number of klukes would decrease, but snive and vortek populations would increase. The joons would then decrease after exhausting much of their food supply. At this point, the number of klukes would increase, and population of snive and vortek would decrease.

8. Energy is lost at each trophic level; the pyramid shape reflects the decrease in numbers of organisms which can be supported by each successive trophic level.

Follow-up Activities

● Have students draw food pyramids as they did in the activity, using data—including food chain members and numbers of each—gathered as they observe:

—a small area of land outside;

—an aquarium;

—a film of an unfamiliar ecosystem; or

—a photograph of a food chain.

- Or, have students obtain data on the numbers of organisms in a large ecosystem from local wildlife, forestry, and fishery agencies. Compare changes in the numbers of members over time. Explain these changes.

- Up to this point, the shape referred to as a "food pyramid" represents a pyramid of numbers. A pyramid of biomass can be constructed in the same way. Have students gather organisms from an ecosystem such as an aquarium. Obtain numbers of grams of dry weight of these organisms (biomass) by drying and weighing them, or by consulting references.

  NOTE: Killing vertebrate animals may be disturbing to some students. As organisms in an aquarium die naturally, freeze them and save them for this activity.

- It has been stated that if human beings ate "lower on the food chain," there would be enough food to feed all of the hungry people in the world. After reading about this topic have students:

  —explain the rationale behind this statement; and

  —argue whether they believe the statement to be true.

- Have students draw the shapes of the food pyramids resulting from each change suggested in question 7 on the student worksheet.

Evaluation

Method of Evaluation	*Goal(s) Tested*
1. Completion of student worksheets	1,2,3
2. Participation in class discussion	1,2,3
3. Evaluation Questions	1,2

Evaluation Question Answers

Provide students with information about actual ecosystems with hypothetical numbers (given below) and have them complete a separate Evaluation Questions Worksheet for each ecosystem.

Ecosystem 1—A Forest

In one small area of land were found:

Organism 1: 75 plants, which were consumed by

Organism 2: 20 rabbits, which were preyed upon by

Organism 3: 4 foxes

Ecosystem 2—The Chesapeake Bay Estuary

In one small area of estuary were found:

Organism 1: 200 marine plants, which were consumed by

Organism 2: 100 crustacea, which were preyed upon by

Organism 3:   60 sea nettles, which were consumed by

Organism 4:   15 striped bass

Ecosystem 3—An Island in the Galápagos

In one small area of the island were found:

Organism 1: 50 cacti, which were consumed by

Organism 2:   5 land iguanas, which were preyed on by

Organism 3: 25 ticks (5 per iguana), which were consumed by

Organism 4:   5 finches

References

1. White, A., Epler, B., & Gilbert, C., *Galapagos Guide*. Quito, Ecuador: Imprenta Mariscal, 1985.

2. Wright, Emmett L., *Chesapeake Bay Environmental Educational Project*. College Park, Maryland: Science Teaching Center, University of Maryland, 1980.

Information about the Chesapeake Bay food chain is from *Decision Making*: *The Chesapeake Bay*, copyright 1985 by the University of Maryland Sea Grant College.

© 1991 by The Center for Applied Research in Education

Name _____   Date _____

WHAT IS A FOOD PYRAMID?

Background

Biologists in the year 2014 observe a food chain which is part of an ecosystem on a distant planet. They find that it has many characteristics of food chains on earth. Here is what they discover:

On a particular area of land, a small, plant-like organism capable of photosynthesis, called a vortek, serves as a source of food for an animal called a snive. The snive, in turn, is preyed upon by a kluke. There are 100 vorteks, 50 snives, and 25 klukes. The food chain may be diagrammed as shown in Fig. 7–2.

FIGURE 7–2. SAMPLE FOOD CHAIN.

Building a Food Chain

By completing the activity described below, you will better understand interactions occurring in food chains.

1. On a blank sheet of paper, list the members of this food chain by:
   a. placing them in order, with the members at the bottom of the food chain near the bottom of the page; and
   b. spacing them 5 cm apart.
2. For each, beginning with the bottom one, draw a horizontal block around it to represent the numbers of each member inside the block following these guidelines:
   a. *height*: Each block should be 5 cm tall, so that it touches the bottom of the block above it.
   b. *width*: Make the width to scale according to the numbers of each organism in the food chain—10 organisms in 1 cm. (The block around the vorteks, for example, will be 10 cm wide.)
   c. *centering*: Center each block on the page.

WHAT IS A FOOD PYRAMID? (continued)

3. Draw lines to connect the edges of the blocks. What shape is formed?

_____

4. Each block represents a trophic level. Label the organisms in each trophic level as to which is a:

   a. producer   _____

                      _____

   b. primary consumer   _____

                      _____

   c. secondary consumer   _____

                      _____

   Explain in the spaces above why you chose these labels for each one.

5. This food chain is similar to many on earth. What factors do you think cause certain food chains to have this shape?

   _____

   _____

6. Turn your drawing upside down. Describe a food chain on earth which might have a shape similar to this one.

   _____

   _____

7. How would the numbers of each member of the food chain—vorteks, snives, and klukes—change over the course of time in each situation below? Assume there are no other predators or food sources unless otherwise indicated.
   a. the biologists visiting the planet consume all of the vorteks for food;

   _____

   _____

   b. a deadly disease wipes out all of the klukes;

   _____

   _____

WHAT IS A FOOD PYRAMID? (continued)

c. animals called skunkolas travel to the area being studied from another region of the planet. The skunkolas prey upon the vorteks; the klukes, however, dislike the taste of the skunkolas and refuse to eat them.

_____

_____

_____

d. the biologists remove all the snives for study;

_____

_____

_____

e. more vorteks are planted;

_____

_____

_____

f. a tertiary consumer called a joon is introduced.

_____

_____

_____

8. It has been stated that it is more energy efficient to eat "lower on the food chain." Look again at the shape of the food chain you drew and explain why this statement is true.

_____

_____

_____

_____

Name _____ Date _____

FOOD PYRAMID
EVALUATION QUESTIONS

Your teacher will provide you with information about actual ecosystems. For each, complete the following steps:

1. Draw the shape of the food pyramid, and describe the reasons for this shape. Reasons:

   _____

   _____

   _____

   _____

   _____

2. Which members are producers and primary, secondary, and tertiary consumers?

   _____

   _____

   _____

3. Complete the table below, indicating the effect on each member in the horizontal column of removing each member in the vertical column.

	effect on:			
	ORGANISM 1	ORGANISM 2	ORGANISM 3	ORGANISM 4
removed:				
ORGANISM 1				
ORGANISM 2				
ORGANISM 3				
ORGANISM 4				

ACTIVITY 8

WHAT ARE THE EFFECTS OF HUMAN INTERVENTION ON A FOOD WEB?

Goals

After completing this activity, students will:

1. understand that making even one change in an ecosystem can affect it profoundly; and
2. be aware of the dangers of human intervention in the environment.

Synopsis

Students will be presented with a food web into which DDT was introduced by the World Health Organization in the 1950s. They will predict the consequences of pesticide introduction into this food web and propose a solution. Their ideas will then be compared with the actual events.

Relevant Topics

food webs
ecosystems
predator-prey relationships
environmental science
factors affecting population growth

Age/Ability levels

grades 7–12, most ability levels

Entry Skills and Knowledge

Before participating in this activity, students should be able to:

1. define the term food web; and
2. describe general aspects of predator-prey relationships.

Materials

photocopied student worksheets

DIRECTIONS FOR TEACHERS

Preparation

1. photocopies of student worksheet: "What Are the Effects of Human Intervention on a Food Web?"

Background

1. The results of spraying DDT on Borneo were that:
   - The geckoes, small lizard-like animals, received nerve damage from eating the cockroaches, and their reflexes were slowed by the DDT.

- Cats were able to catch and prey upon more of them than before; cats consequently died from DDT ingestion.
- Because the cats died, rats, having no predators, moved in from the forest. They carried fleas which carried the plague, and humans contracted the plague.
- When geckoes died due to DDT poison and being consumed by cats, the caterpillar population increased due to a lack of predators.
- Since the increase in numbers of caterpillars meant that more of the thatched hut roofs were consumed, the roofs collapsed.

2. The World Health Organization (WHO) solution was to parachute in more cats to Borneo.[1]

Teaching the Activity

1. Distribute student worksheets, and explain the assignment briefly to students.
2. It is your decision whether to have students work individually or in pairs. Allow time in class for them to complete their worksheets.
3. Provide help as needed. Questions which will assist students in getting started include the following:
   - If the geckoes consumed the cockroaches, what would happen to the geckoes?
   - What animal preys upon the gecko? What is the likely consequence to this predator?
   - What animal does the gecko prey upon? What is the likely consequence to the gecko?
4. Assemble students as a group. Call for volunteers to list some of the consequences and solutions on the board. Accept all reasonable ideas.
5. Explain what really happened in Borneo. You may want to copy the teacher background section to distribute to students. Compare student predictions and solutions with those of the WHO.

Answers to Questions on Student Worksheets

Writing Your ideas

1a. through 1e. answers will vary
2. answers will vary

Follow-up Activities

- Have students read about current examples of human intervention in ecosystems for which the results and solutions are still uncertain. Students should predict what they think will be the effects of this intervention and propose solutions. Some examples include the following:
  —the effects of acid rain on plant and animal life;
  —the effects of human population growth on organisms which populate the Chesapeake Bay;
  —the effects of tourism on species endemic to the Galápagos Islands.

---

[1] Excerpts regarding the Borneo food chain adapted from *Biology*, second edition by Karen Arms and Pamela S. Camp, copyright © 1982 by Saunders College Publishing, a division of Holt, Rinehart and Winston, Inc., reprinted by permission of the publisher.

● Have students create hypothetical food chains on distant planets, and predict the consequences of intervention by Earthlings.

Evaluation

Method of Evaluation	Goal(s) Tested
1. Completion of student worksheets	1,2
2. Participation in class discussions	1,2
3. Evaluation Question 1	1,2

Evaluation Question 1

Provide students with information about members of a specific food web and their interactions; see "What Is A Food Pyramid?" for some examples. Ask them to predict the effect of the removal of every other member on each member of the food web.

Reference

Arms, K., and Camp. P. S., *Biology* (2nd ed.). Philadelphia: Saunders College Publishing, 1982.

WHAT ARE THE EFFECTS OF HUMAN INTERVENTION ON A FOOD WEB?

In the 1950s, malaria was a severe problem in Borneo. Since malaria is carried by mosquitoes, the World Health Organization (WHO) attempted to rid the island of mosquitoes by spraying the pesticide DDT.

The diagram below illustrates a food web in the ecosystem whose members were affected by the DDT. You will predict the effects of the introduction of DDT on each member of the food web, and propose a solution. Your ideas and those of your classmates will then be compared with the actual events.

Consider the following facts as you study the food web shown:

1. DDT causes nerve damage, slow reflexes, and eventual (but not usually sudden) death in an animal.

2. When an animal preys upon another animal which has consumed DDT, the predator is affected by DDT, as well as the prey.

3. In addition to mosquitoes, cockroaches are also affected, but not killed, by DDT.

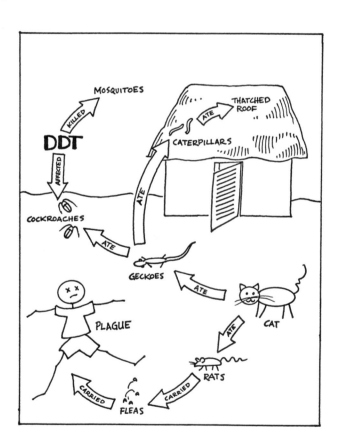

© 1991 by The Center for Applied Research in Education

WHAT ARE THE EFFECTS OF HUMAN INTERVENTION ON A FOOD WEB (continued)

Writing Your Ideas

1. Beginning with the geckoes, list each member of the food web. Beside each, explain the probable effect of DDT on it. Support your answers with evidence from the diagram and the facts you were given about the effects of DDT on organisms in the food web.

a. _____

b. _____

c. _____

d. _____

e. _____

2. What is your solution to this problem? Explain why you think your solution will work, supporting your idea with evidence from the diagram and the facts you were given about the effects of DDT on the organisms in the food web. Then wait for further directions from your teacher.

_____

_____

_____

_____

ACTIVITY 9 _____

IS THE WATER AROUND YOU SAFE TO DRINK?

Goals
After completing this activity, students will:

1. be familiar with sources of bacterial contamination in water;
2. understand the need for purification of drinking water;
3. be aware of the quality of familiar sources of water; and
4. be able to test for the presence of coliform bacteria in water.

Synopsis
Students will make predictions about whether or not water from various sources is safe to drink. They will then test the water samples for the presence of coliform bacteria. If time permits, findings will be presented at a mock city council meeting.

Relevant Topics
environmental pollution
water purification
bacteriology

Age/Ability Levels
grades 10–12, average to gifted

Entry Skills and Knowledge
Before participating in this activity, students should be able to:

1. manipulate bacteria safely in the lab; and
2. define bacteria as microscopic organisms, some of which are harmful to humans.

Materials
photocopies of student worksheets: "Is the Water Around you Safe to Drink?"
"Data Sheet for Presumptive Coliform Test"
"Data Sheet for Confirmed Coliform Test"

per pair of students
one empty Petri dish
one Bunsen burner
two pairs safety goggles
two lab aprons
three sterile sampling bottles with screw top lids (make sure they can be autoclaved safely)
four EMB agar Petri dishes
ten phenol red-lactose broth tubes
fourteen sterile 1 ml pipettes (plugged with nonabsorbent cotton) with bulb
one small glass rod spreader (or a long Pasteur pipette and Bunsen burner)
one test tube rack
one marking pencil
one test tube containing sterile distilled water

per class
one incubation oven set at 37°C
disinfectant
matches
ethanol
parafilm

DIRECTIONS FOR TEACHERS

Background Information

Many diseases are spread through water contaminated with fecal matter. To avoid testing for every type of bacterium, the presence of indicator organisms found only in fecal matter is determined. The most common of these is *Escherichia coli.*

Preparation

1. Sterilize empty, clean water sample bottles in an autoclave or pressure cooker. Lids should only be on "finger tight" during the heating process.

2. Prepare and sterilize test tubes with screw tops or nonabsorbent cotton plugs which are partially filled with distilled water.

   Note: It is your decision whether to let students make up their own tubes and agar plates. Phenol red–lactose fermentation broth and EMB agar may be purchased commercially. Prepare them according to directions, adding 0.5% lactose to the EMB agar, or prepare phenol red–lactose broth tubes by following these steps:*

   ● Combine in 1 l of distilled water:

     —1g Bacto-Beef Extract

     —10g Proteose Peptone No. 3, Difco

     —5g sodium chloride

     —0.018g Bacto-Phenol Red

     —10g (1%) lactose

   ● Adjust pH to 7.4 ± 0.2

   ● Make broth tubes by pouring the broth into a large test tube, inverting it into the larger one. Remove bubbles.

   ● Stopper the tubes and autoclave.

   ● Cool tube contents before inoculating them.

   To prepare EMB (eosin-methylene blue)-lactose agar plates, follow these steps:*

   ● To a small amount of distilled water, add:

     —10g Bacto Peptone

     —5g Bacto Lactose

     —5g Bacto Sucrose

     —2g Dipotassium Phosphate

     —13.5g Bacto Agar

---

* EMB agar and phenol red–lactose broth recipes © *Difco Manual of Dehydrated Culture Media and Reagents for Microbiological and Clinical Laboratory Procedures*, 10th ed. Difco Laboratories, Inc.: Detroit, Michigan (1984). Used by permission.

—0.4g Bacto Eosin Y

—0.065g Bacto Methylene Blue

- Make up to 1 l with distilled water.
- Adjust pH to 7.2 ± 0.2
- Autoclave.
- Pour agar into sterile Petrie dishes on a disinfected surface.

You can also make glass rod spreaders, if you do not have them, in this way: heat the straight end of a long Pasteur pipette and bend it so that it is at a 90-degree angle from the rest of the pipette. Heat the tip and bend it up, as shown in Fig. 9–1.

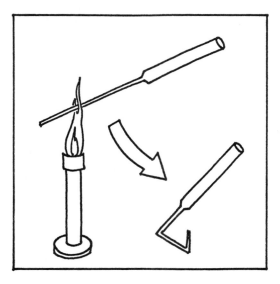

FIGURE 9–1

3. Obtain all student materials.

4. Photocopy student lab and data sheets.

SAFETY NOTES:

- Exercise caution when making glass spreaders. Do not touch them when they are hot. Do not touch the source of heat.
- Goggles should be worn by students and teachers at all times when a flame is being used in the laboratory.
- *E. coli* is pathogenic; GREAT care should therefore be exercised to avoid contamination. Sterilize the work area thoroughly with disinfectant before and after using bacteria. Sterilize Petri dishes, pipettes, test tubes, and liquids which come into contact with bacteria by autoclaving. Sterilize pipettes which cannot be autoclaved with a strong disinfectant. Never allow students to mouth pipette solutions containing bacteria; all pipetting should be done with a bulb. Have students wash their hands at the beginning and at the end of the lab period. Since other pathogens might grow at 37°C, be especially careful not to spill tubes or Petri dishes inoculated with *E. coli*. Wrap Petri dishes with parafilm before incubating them to prevent spills.

- Follow the directions in the owner's manual for proper use and care of the autoclave.
- Set up an area in the laboratory, such as a cart covered with aluminum foil, on which to place materials needing to be sterilized, such as pipettes. Instruct students to place contaminated equipment here.
- As with all laboratories, remind students to exercise caution in handling all laboratory chemicals. Do not ingest or inhale them directly and avoid contact with eyes, clothing, and skin. This is particularly important in this lab, since *E. coli* is a human pathogen. Flood with water immediately if contact occurs.
- Check local and county ordinances regarding disposal of autoclaved culture media.
- If laboratory spills occur, do the following:

  —if a fire hazard exists, extinguish all flames immediately:

  —dilute the spill with water;

  —apply a commercial absorbent such as kitty litter to the spill; and

  —clean the area thoroughly with soap and water, then mop it dry. Wear rubber gloves, and use a dustpan and brush while cleaning up.

- Have students wear laboratory aprons at all times.
- Allow plenty of time for students to clean up at the end of each lab period to reduce the chances of laboratory accidents. Have them clean their work areas thoroughly at the end of each lab.
- Warn students of specific safety precautions regarding these chemicals:

  —*stains*: (methylene blue and phenol red in the media) Avoid contact with skin, clothing, and eyes, which can be stained by these substances.

  —*ethanol*: Ethanol is flammable, so GREAT care should be taken when using it to sterilize the glass spreader. Pour a small amount into a sterile glass Petri dish. Keep the dish covered when not dipping the spreader. Keep the stock container of ethanol far away from the open flame. Make sure students wear goggles and a lab apron at all times they are working with the ethanol.

Teaching the Activity

1. Explain safety and background information to students.
2. Distribute student laboratory sheets, explaining the assignment briefly.
3. Have students work in groups of two. Allow them time in class to choose three sources of water to sample and to write hypotheses.

   If students have difficulty listing sources of water, here are some suggestions:

   - a city water reservoir
   - a saline treatment plant
   - a lake, pond, river, stream, or ocean
   - a mud puddle
   - a well
   - different areas of the same large water source
   - water faucets at home or school
   - public drinking fountains or
   - water from a toilet bowl (before and after cleaning)

4. Write your approval on the planning sheets. You may want to complete this step overnight and return the planning sheets to students on the following day.

5. Instruct students on how to take a water sample by following these steps:*

   From a moving body of water: Hold the open sterile container into the flow of water so that water from the hands does not contaminate the sample.

   From a faucet: Allow the water to run for at least ten minutes prior to taking a sample; do not allow the container to come in contact with the faucet.

6. Allow students to take containers home with them overnight to obtain samples. On the following day, perform the laboratory tests.

7. Make observations after 24 hours. Have students record data on the chalkboard. Be sure they have completed their laboratory sheets.

8. Bring students together as a group. Ask questions such as those below to stimulate discussion:

   • How many hypotheses were supported? How many refuted?

   • What were your hypotheses and reasons for them?

   • If your hypothesis was refuted, how might you explain this result?

   • What might be the sources of contamination in the water for which you obtained positive tests?

   • Why does water need to be cleaned before and after it is used by humans?

   • How might the sources of contamination be eliminated?

9. If there is time, have students carry out one of the follow-up activites.

Answers to Questions on Student Worksheets

Preparation

1. and 2. Answers will vary.

3. answers will vary. A sample hypothesis and rationale might be: There will not be any coliform bacteria in the water from the lake near our school. The rationale for this is that many plants and animals live in that water; it seems that they would die if the water were contaminated.

Questions

1. through 3. Answers will vary.

Follow-up Activities

• If evidence of fecal contamination is found in a public body of water as a result of this laboratory, (or even if not, just pretending it was found), have students prepare a case to present to the members of the city council regarding the problem. Points which should be addressed include:

   —why research should be done to determine the source of the contamination;

   —why the bacteria should be eliminated; and

   —how they can be eliminated.

---

* (Sampling procedures from Brock/Brock, *Basic Microbiology With Applications*, © 1978, pp. 428, 429. Adapted by permission of Prentice-Hall, Inc., Englewood Cliffs, N.J.)

- Have students present the cases to the class in a mock city council meeting. Other members of the class can play city council members who decide how to deal with the problem based on the evidence. Considerations of city council members include:

  —how their decisions will affect their reelection campaigns;

  —how much the changes will cost;

  —the sources of funds for the changes;

  —who will implement the changes;

  —how the public will be made aware of the problem; and

  —how the public will react to the situation.

- If a serious public water contamination problem is found, consider having the students present the case at an actual city council meeting.

- Visit a water treatment plant.

Evaluation

Method of Evaluation	Goal(s) tested
1. Completion of student laboratory sheets.	1,2,3,4
2. Participation in class discussions.	1,2,3,4
3. Lab exercise 1.	1,4

Lab Exercise 1

Provide students with two water samples to test—one contaminated with *E. Coli*, and one not contaminated. Do not indicate which is which. Have students test the samples to determine which has coliform bacteria, and which does not.

References:

1. Brock, T. D., & Brock, K. M., *Basic Microbiology* (2nd ed.). Englewood Cliffs, N.J.: Prentice-Hall, Inc., 1978. (Information about presumptive and confirmed tests used by permission.)

2. *Difco Manual of Dehydrated Culture Media and Reagents for Microbiological Laboratory Procedures*, 10th edition, Difco Laboratories, Inc., Detroit, Michigan (1984). (Recipes used by permission.)

Name _____   Date _____

IS THE WATER AROUND YOU SAFE TO DRINK?

You are going to form and test hypotheses about whether water from three different sources is contaminated with fecal matter. Your teacher will provide you with background information about coliform bacteria.

Preparation

1. List three different sources of water that you wish to test for bacterial contamination.

   a. _____

   b. _____

   c. _____

2. For each source, indicate in the space above whether you expect to find fecal contamination there.

3. For each water source, write a hypothesis about whether coliform bacteria will be present. Underneath each, write a one-sentence rationale for it.

   Hypothesis #1: _____

   _____

   _____

   Hypothesis #2: _____

   _____

   _____

   Hypothesis #3: _____

   _____

   _____

4. Have your teacher approve your work up to this point.

Teacher Approval _____

5. On your data sheets, record your choices and expected results.

SAFETY NOTES:

- *E. coli* is a human pathogen, so GREAT care must be exercised in avoiding contamination. Follow all safety precautions as directed by your teacher, including those listed here. Sterilize the work area thoroughly with disinfectant before and after using bacteria. Never mouth pipette solutions containing bacteria; all pipetting should be done with a bulb. Wash your hands at the beginning and at the end of the lab period. Place materials which come into contact with bacteria in a specific area of the lab, as directed by your teacher. Follow procedures specified in the lab very carefully.

- Wear goggles and a lab apron at all times when a flame is being used in the laboratory.

- As with all laboratories, exercise caution in handling all laboratory chemicals. Do not ingest or inhale them directly, and avoid contact with eyes, clothing, and skin. Flood with water immediately if contact occurs.

IS THE WATER AROUND YOU SAFE TO DRINK? (Continued)

- If laboratory spills occur, inform your teacher at once.
- Exercise caution when using these chemicals:

  —*stains* (methylene blue and brom cresol purple in the media) Avoid contact with skin, clothing, and eyes, which can be stained by these substances.

  —*ethanol*: Ethanol is flammable, so GREAT care should be taken when using it to sterilize the glass spreader. Pour a small amount into a sterile glass Petri dish. Keep the dish covered when not dipping the spreader. Keep the stock container of ethanol far away from the open flame.

Performing the Experiment

You will perform two laboratory tests; directions follow. The lactose fermentation test, called the "presumptive" test, is usually performed first by health departments testing water. If a positive result is found, a "confirmatory" test is then carried out on EMB-lactose agar. To save time, you will perform both tests on the same day. When you gather all of your data, answer the questions at the bottom of the data sheet for the confirmed test.

Phenol Red–Lactose Test (Presumptive Test)

1. Phenol red–lactose tubes have been prepared for you. Obtain 10.
2. Label each tube, including its contents and your initials. Perform each test sample in triplicate; the tenth tube is the control.
3. Transfer 1 ml of test water into each tube except the control, swirling the test substance first.

   SAFETY NOTE: Transfer all test substances with a bulb. Never mouth pipette solutions that are known or suspected of containing bacteria.

4. Transfer 1 ml of sterile distilled water into the control tube.
5. Swirl the tubes well to mix contents. Do not introduce air bubbles into the small tube.
6. Return materials to the proper area, as designated by your teacher.
7. Incubate tubes at 37°C for 24 hours.
8. Observe tubes for the presence of gas bubbles in the smaller tube, as shown in Figure 9–2. Also

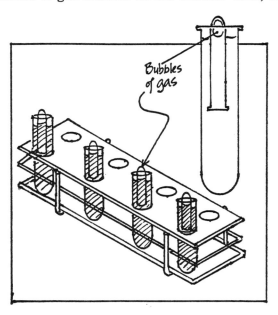

Bubbles of gas

FIGURE 9–2

IS THE WATER AROUND YOU SAFE TO DRINK? (Continued)

observe to see whether the color has changed from red to yellow. Indicate in Table 9–1 of your data sheet for the presumptive test the results, using a "+" or "−" for each tube.

9. Record your data on the chalkboard.

EMB Agar Test (Confirmed Test)

1. Obtain four EMB agar plates; the fourth one is the control.
2. Label each on the BOTTOM; include contents, date, and your initials.
3. Transfer 1 ml of each test substance onto the surface of three of the plates. Transfer 1 ml of sterile distilled water onto the control plate.

   SAFETY NOTE: Transfer all test substances with a bulb. Never mouth pipette solutions containing bacteria.

4. Dip the glass spreader into ethanol which has been poured into an empty Petri dish.

   SAFETY NOTE: Wear goggles and a lab apron whenever you are working with ethanol, and follow safety precautions for flammable materials.

5. Flame the spreader briefly, remove from flame.
6. When the ethanol has burned off (when the blue flame on the glass rod is gone), discharge the heat by touching it briefly to the surface of the agar plate where water has not been transferred.

   SAFETY NOTE: Keep the lid on the Petri dish at all times you are not transferring or spreading a substance onto the surface of the agar.

7. Spread the water sample around the plate with the glass spreader, as shown in Figure 9–3.

Dip the pipette in ethanol, then flame briefly. Move the flat edge over the surface of the agar.

FIGURE 9–3

8. Dip the spreader in ethanol and flame it again.
9. Repeat steps 4 through 8 for each sample.
10. Incubate plates UPSIDE DOWN at 37°C for 24 hours.
11. Observe the surface of the agar. If *E. coli* is present, colonies will have a green metallic sheen. Record your results on the data sheet for the confirmed test in Table 9–2.
12. Record your results on the chalkboard.

DATA SHEET FOR PRESUMPTIVE COLIFORM TEST

Table 9–1

Source of Water	Tube #	Expected Results	Actual Results (Presumptive)
Experimentals:			
1	1		
	2		
	3		
2	1		
	2		
	3		
3	1		
	2		
	3		
Control:			

DATA SHEET FOR CONFIRMED COLIFORM TEST

Table 9–2

Source of Water	Plate #	Expected Results	Actual Results (Confirmed)
Experimentals: 1	1		
2	2		
3	3		
Control	4		

Questions

Complete the information below after you have gathered your data.

1. Which of your samples are safe to drink? Which ones are not? _____

_____

_____

2. Which of your hypotheses were supported? Which were refuted? _____

_____

_____

3. If any of your samples were contaminated, list possible sources of contamination. _____

_____

_____

_____

ACTIVITY 10

HOW IS AN ORGANISM ADAPTED TO ITS ENVIRONMENT?

Goals

After completing this activity, students will:

1. be better observers of organisms in their natural habitats;
2. better understand life functions of organisms; and
3. understand how organisms are adapted to their environments.

Synopsis

Students will observe organisms in their environments, then attempt to discover examples of adaptation in these organisms.

Relevant Topics

symbiosis
evolution
adaptation
plant, animal, and fungus:
   taxonomy
   reproduction
   methods of obtaining food

Age/Ability Levels

grades 7–12, most ability levels

Entry Skills and Knowledge

Before participating in this activity, students should be able to:

1. use a taxonomic identification key correctly, and
2. define the terms *symbiosis, parasite, saprophyte*.

Materials

student worksheets
drawing pencils, crayons, or magic markers

DIRECTIONS FOR TEACHERS

Preparation

1. Photocopy student worksheets: "Animal Observation," "Fungus Observations," "Plant Observation"
2. Gather drawing pencils, paper, and crayons (or require students to obtain their own).

SAFETY NOTES:

- Warn students as they make their observations to stay away from animals such as insects which may sting or bite. Know ahead of time if individual students have dangerous allergies; if so, carry appropriate remedies with you.
- Remind students to keep their hands away from their mouths when handling plants and fungi. Some species are poisonous. Have them wash their hands immediately after each observation. Warn them not to touch plants which can cause allergic reactions, such as poison ivy.
- Be prepared to answer questions from students about which organisms are potentially harmful. Consult your state or local safety manual for a list of these. With students, it is probably wise to assume *all* species of mushrooms are poisonous, even though they are not.

Teaching the Activity

There are several ways to implement this exercise. Two are described below:

A. Use the separate observations (fungus, plant, and animal) to introduce units on each type of organism. The questions on the worksheets that students are unable to answer can form the basis for what they are to learn. Repeat the observation after students have studied about the organisms. Comparison of pre- and postobservation worksheets will provide an assessment of how much students have learned.
B. Use the observations at the end of units as reviews.

The following directions are written for method A.

1. Choose an area in which students are likely to find an abundance of the organisms they will observe. Here are some suggestions:

   fungi: a wooded area, preferably after a rainstorm

   plants: an outdoors area or a botanical garden

   animals: a wooded area, an aquarium, or a zoo
2. Distribute student worksheets, and briefly explain the assignment. Be sure students have drawing pencils, paper, and crayons.
3. In the "field," be sure the organism each student is observing is truly an example of the type he or she has decided to study. If not, direct the student to a correct example.
4. Allow time for observations—at least 30 minutes, longer if possible.
5. On the following day, have students use taxonomic keys to identify organisms.
6. Share observations as a class, including drawings. Ask students
   - what questions they were able to answer;
   - which ones they had difficulty answering; and
   - what new questions they wrote.
7. Use the student responses as a basis for study, having them do library research to answer all questions in the unit with which they had difficulty. Answers will vary to all questions on student worksheets, depending on the organisms observed.
8. Repeat the observation process on the same organism in the same habitat, if possible, at the end of the unit.

9. Have students compare their answers to questions during the two observations. Ask them to share what they learned and any new questions they have.

Follow-up Activities

- Have students observe the same organisms at different times of the year. Compare the answers to determine whether they change with time.
- Have students watch videotapes of or read about scientists such as Jane Goodall to learn more about techniques of field observation of living organisms.
- Show slides and videotapes of examples of unusual ways organisms have adapted to environments on earth.

Evaluation

Method of Evaluation	Goal(s) Tested
1. Completion of student observation sheets	1,2,3
2. Participation in class discussion	1,2,3
3. Evaluation Questions worksheet	1,2,3

Answers to Evaluation Questions:

Answers will vary; look for evidence of growth in student understanding.

Name _____ Date _____

ANIMAL OBSERVATION

You will observe an animal in its natural habitat, then attempt to discover examples of adaptation in this animal. Answer the questions and make a drawing of your animal. Be prepared to share your findings with the rest of the class.

SAFETY NOTE:

As you make your observations, stay away from animals such as insects which may sting or bite. If you are in doubt about whether an animal is harmful, ask your teacher.

1. Where does your animal live (in soil, in a tree, under a rock)? _____

    _____

2. Does it live alone or with others? _____

    _____

3. Describe the plants around it. _____

    _____

4. Describe the animals around it. _____

    _____

5. Does there appear to be competition between your animal and others for food, living space, or water? If not, how might this animal be adapted so as to minimize competition for these factors?

    _____

    _____

6. Do you observe any symbiotic relationships? If so, describe these. _____

    _____

    _____

7. How does the animal appear to:

    a. Obtain food (what type of food)? _____

    _____

    b. Protect or defend itself? _____

    _____

    c. Reproduce? _____

    _____

ANIMAL OBSERVATION (continued)

d. Communicate? _____

_____

e. Sense its environment? _____

_____

8. Does it have a long learning period with its parents? _____

_____

9. Does it have any predators? What are they? _____

_____

10. Classify your animal as to its:

a. kingdom _____

b. phylum _____

c. (subphylum) _____

d. class _____

e. order _____

f. family _____

g. genus _____

h. species _____

11. Describe any unusual features. _____

_____

12. On a separate page, draw a picture of the animal in its surroundings.

13. List any questions you have about your animal not asked so far. _____

_____

_____

14. Based on your observations, list ways in which the animal is adapted to its environment.

_____

_____

_____

Name _____ Date _____

FUNGUS OBSERVATION

You will observe a fungus in its natural habitat, then attempt to discover examples of adaptation in this organism. Answer the questions and make a drawing of your fungus. Be prepared to share your findings with the rest of the class.

SAFETY NOTE:

Keep your hands away from your mouth when handling fungi. Some species are poisonous. Wash your hands immediately after each observation. If you are in doubt about whether a fungus is harmful, ask your teacher.

1. Where does your organism live (on the ground, on a living tree, on a decaying plant)?

   _____

2. Is it a parasite or saprophyte? How do you know? _____

   _____

3. Describe its surrounding environment. _____

   _____

4. Do you see evidence of a symbiotic relationship of the fungus with another organism? If so, describe it. _____

   _____

   _____

5. Does it appear to need sunlight directly to live? How do you know? If direct sunlight is not necessary, how does it obtain its food? _____

   _____

6. What is its source of water? _____

   _____

7. How do you think it reproduces? Do you see any evidence to support your idea? _____

   _____

   _____

8. Do you see any animals around it? If so, describe them. _____

   _____

FUNGUS OBSERVATION (continued)

9. Does it appear to have any predators? If so, how might it be protected against them?

_____

10. What do you consider to be its value to its surrounding environment? _____

_____

11. To what class does it belong: Oomycetes, Zygomycetes, Ascomycetes, Basidiomycetes, or Fungi Imperfecti? How do you know? _____

_____

_____

12. Describe any unusual features. _____

_____

13. On a separate page, draw a picture of the fungus in its surroundings.

14. List any questions you think of about the fungus not already listed. _____

_____

_____

15. Based on your observations, list ways in which the fungus is adapted to its environment.

_____

_____

_____

_____

_____

Name _____ Date _____

PLANT OBSERVATION

You will observe a plant in its natural habitat, then attempt to discover examples of adaptation in this plant. Answer the questions and make a drawing of your organism. Be prepared to share your findings with the rest of the class.

SAFETY NOTE:

Keep your hands away from your mouth when handling plants. Some species are poisonous. Do not touch plants which can cause allergic reactions, such as poison ivy. When in doubt about whether a plant is harmful, ask your teacher. Wash your hands immediately after each observation.

1. Does your plant grow in soil? If so, describe the soil texture and color. _____

_____

2. What other types of plants live nearby? _____

_____

_____

3. Is it found in the shade of another plant? If so, how is it adapted to the shade? _____

_____

4. Do you see evidence of symbiosis around the plant? If so, describe it. _____

_____

5. Describe and draw the arrangement of veins in the leaves.

_____

_____

6. Do the leaves have any special adaptations to the environment? Stems? If so, describe these. _____

_____

7. What kinds of roots, if any, is this plant likely to have—deep or shallow? Why? _____

_____

8. How do you think this plant reproduces? Do you see any evidence to support this idea?

_____

_____

PLANT OBSERVATION (continued)

9. Does it require a lot of water? Why or why not? _____

_____

10. Classify your plant as to its:

   a. kingdom _____

   b. phylum _____

   c. (subphylum) _____

   d. class _____

   e. order _____

   f. family _____

   g. genus _____

   h. species _____

11. Describe any unusual features. _____

_____

12. On a separate page, draw a picture of the plant in its surroundings.

13. List any questions about your plant not listed so far. _____

_____

_____

_____

14. Based on your observations, list ways in which the plant is adapted to its environment.

_____

_____

_____

_____

_____

Name _____ Date _____

ORGANISM OBSERVATION
EVALUATION QUESTIONS

1. Compare your observation sheets from before and after research and study.

   a. What answers could not be answered until your research was done? _____
   _____

   b. What answers remained the same? Be specific. _____
   _____

   c. What answers changed? Be specific. _____
   _____

2. What did you learn about the organism? _____
   _____
   _____

3. What did you learn about how organisms adapt to their environments? _____
   _____
   _____

4. What did you learn about the process of observing? _____
   _____
   _____

5. What new questions do you have about the organism? _____
   _____
   _____

6. What other organisms would you be interested in observing? Why? _____
   _____
   _____

ACTIVITY 11

CAN YOU DESIGN A NEW LIFE FORM?

Goals

After completing this activity, students will:

1. understand difficulties involved in defining life; and
2. better understand structural and functional adaptations of an organism to its environment.

Synopsis

As a class, students will decide on criteria to define an organism as living. Each student will then "create" a hypothetical planet and design a life form suited to survive in that environment.

Relevant Topics

evolution
characteristics of life

Age/Ability Levels

grades 7–12, most ability levels

Entry Skills and Knowledge

Materials

photocopies of student worksheet "Can You Design a New Life Form?"
drawing paper
crayons or magic markers
(optional) photocopies of "Designing a Lifeform: Evaluation Questions"

DIRECTIONS FOR TEACHERS

Preparation

1. Photocopy student worksheets.
2. Gather materials for student drawings, or let students buy their own.

Teaching the Activity

I. Defining Life

1. Briefly explain the assignment to students and distribute worksheets. Ask the following questions to stimulate student thought:
   - Must a living creature necessarily:
     —consume oxygen?
     —have a blueprint available, such as DNA, in order to make more copies of itself?
     —be able to adapt to its environment?

95

—be able to reproduce itself, or can something else be responsible for its reproduction?

—be able to grow?

—be able to move?

—require food in order to stay alive?

—be made of matter? If so, must it be carbon based? What are the advantages of a carbon-based existence? What other molecule(s) can undergo the same chemical reactions as carbon?

—be able to convert energy from one form to another? If so, what is the form of this energy?

- Can a living organism be made of energy? What form?

2. Have each student write one characteristic of a living organism, then ask them to share their ideas. Each can list his/her criterion for life on the board, with examples.

3. After many ideas are listed, consider each one separately as a class. Test each criterion by asking whether there is a nonliving object which also fits this criterion. For example:

Movement: A Slinky® can move, "walking" itself downstairs; mercury also appears to move in a Petri dish.

Oxygen consumption: A flame consumes oxygen (and moves).

Reproduction: A virus cannot reproduce without a host cell, but some experts classify it as living.

4. The class may not reach a consensus on the criteria that define life. It may be necessary, therefore, to vote on criteria to be used for purposes of this assignment.

5. Erase those criteria not considered by the class to be unique to living organisms. Have students copy the remaining criteria onto their worksheets.

II. Designing the Planet and Life Form

1. Make certain that students have materials they need in order to illustrate their ideas and that they understand the assignment.

2. Implement this phase in one of these ways, depending on your schedule:
- have students complete the activity in class, in pairs or individually; or
- have students complete the activity at home.

3. Assign a due date for the activity.

4. Have students share their ideas with the class. Each student should:
- show the class the illustration; (Figure 11–1)
- explain the characteristics of the planet, and how the life form is suited to survive on it; and
- explain how the life form meets the criteria for a living organism.

Answers to Questions on Student Worksheets

Preparation

1.a. through f. Answers will be those identified by the class in Part I.

2.a. through f. Answers will vary.

Designing Your Planet and Life Form

1.a. through c. Answers will vary.

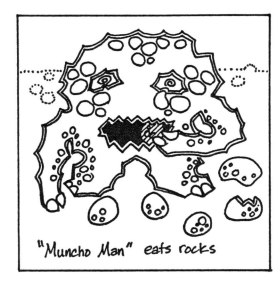

"Muncho Man" eats rocks

FIGURE 11–1

Follow-up Activities

- Have students research the known characteristics of planets and their moons in our solar system. NASA education departments have free materials available on this subject.
- Have students design life forms which might be possible on these planets and moons.
- Show slides and videotapes showing examples of unusual ways organisms have adapted to environments on earth.

Evaluation

Method of Evaluation	Goal(s) Tested
1. Participation in class discussion and sharing sessions.	1,2
2. Completion of student worksheet assignments.	1,2
3. Evaluation Questions	
Question 1	1
Question 2	1
Question 3	2
Question 4	2

Question 1 Answers: will vary.

Question 2 Answers: 2: a. N b. Y c. Y d. N

Question 3 Answers: 3: a. T b. F c. T

Question 4 Answers:

NOTE: Answers may vary from those given below; use your judgment in deciding whether student answers are correct.

a. thick skin which conserves water
b. strong sense of smell, paws which act like shovels
c. yellow flowers and/or reproductive organs
d. white fur, thick layer of fat
e. paws, "fingers," and/or tail for grasping
f. large eyes

CAN YOU DESIGN A NEW LIFE FORM?

You and your classmates will decide on criteria to define life. Using these criteria, you will then "create" a hypothetical planet and design a life form suited to survive in that environment.

Preparation

1. In the spaces below, record the criteria you and your classmates have decided on to define life. Continue on the back of this page if you need to.

   a. _____

   b. _____

   c. _____

   d. _____

   e. _____

   f. _____

2. Complete the information below about your organism and its environment:

   a. Does the organism need to exchange chemicals with its environment? If so, describe these and their functions. _____

   _____

   _____

   b. Describe the organism's food source, if it requires one. Is the food source likely to disappear soon? Explain. _____

   _____

   _____

   c. Describe the climate of the planet, including its temperature, energy, composition, and state(s) of matter present. _____

   _____

   _____

   d. Does your organism have any predators? If so, what are they? _____

   _____

   _____

CAN YOU DESIGN A NEW LIFE FORM? (continued)

e. Is the planet's climate likely to change soon? Explain. _____

_____

_____

f. How large is the planet, in comparison to earth? _____

_____

Designing Your Planet and Life Form

1. Using the information in the preparation section above, complete the following on your own paper, to be handed in:
   a. Write characteristics of the hypothetical planet you have "created."
   b. Describe the organism you have designed. Make certain that it fits the criteria of a living organism and that it is entirely suited to dwelling on the planet you have created. These are the only requirements.
   c. Illustrate your organism in color, showing it in its environment. This may be a poster, drawing, or computer graphic.
2. Be prepared to share your "creation" with the class.
   DUE DATE: _____

DESIGNING A LIFE FORM: EVALUATION QUESTIONS

1. On a separate sheet of paper, list characteristics which define an organism as living. Besides each, indicate:
   a. an example of a living organism with that characteristic; and
   b. a possible argument against that characteristic being a defining criterion of life.

2. For each statement below, indicate in the space provided whether it is an argument for or against viruses being classified as living. Use the following code:
   Y = in favor of
   N = against
   _____ a. Viruses cannot reproduce themselves without a host cell.
   _____ b. A virus can mutate and therefore adapt to its environment.
   _____ c. A virus has its own DNA or RNA and is therefore capable of replicating itself.
   _____ d. Viruses rely on an external source for their energy requirements.

3. The size of appendages, such as ears and tails, of an animal can often be correlated with the climate in which it lives. Indicate which of the statements below is true about this phenomenon by writing *T* (true) or *F* (false) in the appropriate space.
   _____ a. One explanation for this has to do with the amount of heat lost over a given surface area in a particular appendage.
   _____ b. Small appendages would be a disadvantage in a cold climate.
   _____ c. An animal in a warm climate would tend to have large, rather than small, ears.

4. In each case below, indicate at least one type of characteristic that enables the plant or animal to be adapted to its environment:

   a. a plant living in a dry, warm climate _____

   _____

   b. an animal that lives below ground and digs through the earth for food _____

   _____

   c. a plant with a pollinator attracted only to the color yellow _____

   _____

   d. an animal that lives in a cold, snowy area _____

   _____

   e. an animal that lives in trees _____

   _____

   f. an animal that preys at night _____

   _____

section 4

Genetics

Most high school teachers agree that genetics is one of the most difficult topics for their students to learn. The difficulty lies in part in the fact that principles of genetics are complex and abstract.

The activities in this section have been designed to make these principles more concrete and accessible to students. They stress the processes of science in various ways, but all are designed to encourage students to process information actively themselves to better understand these principles.

These activities present simple, inexpensive, hands-on methods of rendering the principles of genetics concrete. "How Do Mutations Occur?" has students manipulate bits of paper representing DNA bases to simulate and discover for themselves how point mutations occur.

The activities entitled "How Are Traits Transmitted in Corn Plants?", "What Is the Effect of Radiation on Plants?", and "What Are Some Interactions Between Genes and the Environment?" all make use of genetically inbred seeds. Because the seeds have mutations which are visible in the phenotypes of offspring, they enable students to observe the result of gene action with the naked eye.

In "How Are Traits Transmitted in Corn Plants?", students determine through observation alone the mutations which are present, and their modes of transmission. "What Are Some Interactions Between Genes and the Environment?" asks students to design and implement experiments which test the effects of environmental agents on the phenotypes of certain plants. Students also design and carry out experiments in "What Is the Effect of Radiation on Plants?" using irradiated seeds.

How Should These Activities Be Used

Below are suggestions for integrating these activities into traditional high school biology courses:

1. "How Are Traits Transmitted in Corn Plants?" can be used as a lab in a unit on principles of Mendelian inheritance. It can be followed by "What Is the Effect of Radiation on Plants?" and "How Do Mutations Occur?" to introduce the concept of mutation.

2. "What Are Some Interactions Between Genes and the Environment?" can be integrated either into a unit on genetics or a unit on environmental science.

ACTIVITY 12

HOW ARE TRAITS TRANSMITTED IN CORN PLANTS?

Goals

After completing this activity, students will:

1. be able to determine phenotypic ratios of corn plants carrying certain mutant traits;
2. be able to determine from these ratios how traits involving certain monohybrid and dihybrid crosses are transmitted; and
3. be more adept at working monohybrid and dihybrid problems.

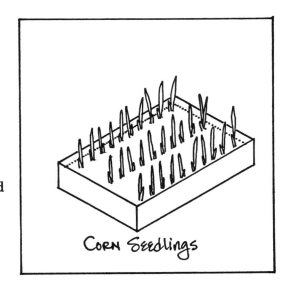
Corn Seedlings

Synopsis

Students will grow seeds of various types of segregating mutant plants and determine phenotypic ratios of certain traits. They will determine the types of mutations present and the genotypes from observing the phenotypes.

Relevant Topics

Mendelian genetics
mutation

Age/Ability Levels

grades 9–12, average to gifted

Entry Skills and Knowledge

Before participating in this lab, students should be able to:

1. describe basic principles of Mendelian genetics;
2. define *phenotype* and *genotype*;
3. work monohybrid problems; and
4. work dihybrid problems (optional).

Materials

photocopies of Student Worksheet: "How Are Traits Transmitted in Corn Plants?"
(optional) photocopies of "Transmission of Traits Evaluation Questions"

per pair of students
two flats, at least 2 square feet in size
vermiculite
potting soil
100 mutant corn seeds
100 normal corn seeds
one marking pencil

per class
labeling tape
watering bottles
plant food

DIRECTIONS FOR TEACHERS

NOTES:

1. Follow this activity with the next one, "What Are Some Interactions Between Genes and the Environment?", if possible.
2. If you are teaching younger or lower ability students, you may modify this lab by:
   - using monohybrid crosses only; or
   - using seeds with only one visible mutation.

Preparation

1. Obtain seeds:

   Mutant seeds:
   Obtain mutant seeds commercially from biological supply houses (The following mutant corn seeds are available from Carolina Biological Supply Company, 2700 York Road, Burlington, N.C., 27215, or Box 187, Gladstone, Ore., 97027 or from a university agricultural department).

   a special 100% dwarf (d1 d1) mutant

   monohybrid mutants:
   green: albino (3:1)
   normal: white tip (3:1)
   tall: dwarf (3:1)

   dihybrid mutants:
   green: albino: red: nonred (9:3:3:1)
   green: albino (9:7)

   Normal seeds:
   Purchase from a seed store, or from a biological supply house.
2. Place mutant seeds in envelopes (100 of each mutant per envelope) with codes which do not indicate the types of mutations present. For example, label envelopes "Mutant A," "Mutant B," etc. Prepare envelopes with at least three different mutants per class.
3. Place normal seeds in envelopes (100 per envelope) labeled "normal corn seeds."
4. Obtain lab materials.
5. Photocopy student worksheets.

SAFETY NOTES:

- If you use any seeds which are treated with pesticides or fungicides, wash them before using.
- Caution students not to eat seeds and to keep all plant parts out of their mouths. Have them wash their hands at the end of each lab period.

Teaching the Lab

1. Distribute student worksheets and briefly explain the assignment.
2. Have students work in pairs. Provide each group with 100 normal seeds and 100 mutant seeds.
3. Demonstrate to students how to prepare seeds in flats.
4. Store flats in a well-lighted area.
5. Allow time every two or three days for students to "feed" their plants.
6. On day 14, allow students at least 30 minutes of class time to make observations and record their data.
7. When all students have completed their worksheets, have each group show the class the visible mutation(s) and compare their observed phenotypic ratios to the "ideal" ratios. Provide them with the commonly used terms for their mutations.
8. Review questions on the worksheets, calling for student responses. Have students work some of their Punnett square problems on the board.

Answers to Questions on Student Worksheet

1. Answers will vary.
2. Plants which have the appearance of the control plants are normal; others are mutants.
3. Answers will vary.
4. See teacher preparation section; answers will vary depending on the mutation the student was given.
5. In some cases, such as normal: albino (3:1), it would be possible to infer this; in others, such as 100% dwarf, no. It would be necessary to breed these plants with normal ones and with themselves, then determine phenotypic ratios of offspring.
6. Answers will vary; for the tall: dwarf (3:1), it might be (if T = tall and t = dwarf):

	T	t
T	TT	Tt
t	Tt	tt

Offspring ratios will be three tall (TT and Tt) and one dwarf (tt)

Follow-up Activities

- Have students perform the experiment which follows this one, "What Are Some Interactions Between Genes and the Environment?".
- Have students repeat this experiment with mutants of other types of seeds, and compare the results with those of corn. Mutant seeds available from Carolina Biological include sorghum, soybean, tobacco, and tomato.
  SAFETY NOTE: Parts of sorghum, tobacco, and tomato plants are toxic. Remind students to keep plant parts away from their mouths and to wash their hands at the end of each lab period.

- Have students simulate genetic crosses involving cat coat color using the computer simulation program CATLAB (available from Conduit). CATLAB allows students to determine how genes for cat coat color are transmitted by observing the phenotypes of offspring of their "crosses."
- Have students simulate point mutations by working through the activity "How Do Mutations Occur?".
- Have students perform the experiment "What Is the Effect of Radiation on Plants?".

Evaluation

Method of Evaluation	Goal(s) Tested
1. Completion of student worksheets and data charts	1,2,3
2. Participation in class discussion	1,2,3
3. Lab activity 1	1,2,3
4. Evaluation Questions worksheet	2,3

Lab Activity 1

Set up a lab station with 100 germinated mutant seeds that are unfamiliar to students and 100 germinated normal seeds of the same species. Ask students to determine from their observations:

a. the type of mutation present;

b. the phenotypic ratio of normal to mutant seeds; and

c. the probable mode of transmission of the mutant gene.

Question 1 Answers: Each of the types listed will have a 25 percent chance of occurring, as shown:

gametes:

$llNn$ $\times$ $Llnn$

lN	Ln
ln	Ln
lN	ln
ln	ln

Punnett Square Cross:

	Ln	Ln	ln	ln
lN	LlNn	LlNn	llNn	llNn
ln	Llnn	Llnn	llnn	llnn
lN	LlNn	LlNn	llNn	llNn
ln	Llnn	Llnn	llnn	llnn

Question 2 Answer: F2 plants include one homozygous tall plant (TT), two heterozygous tall plants (Tt), and one homozygous short plant (tt):

P1 generation = TT × tt

F1 generation =
(all offspring are
Tt, or heterozy-
gous tall)

	T	T
t	Tt	Tt
t	Tt	Tt

F2 generation =
phenotype: 3:1,
Tall: short geno-
type: 1TT: 2 Tt:
1tt

	T	t
T	TT	Tt
t	Tt	tt

Question 3 Answer: The man's genotype is LlGg, and the gametes are shown in Figure 12–1.

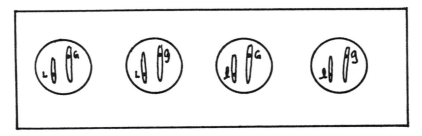

FIGURE 12–1

The woman's genotype is Llgg, and the gametes are shown in Figure 12–2.

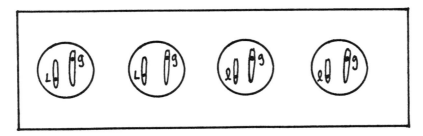

FIGURE 12–2

Reference

Seedlings for Genetics: Segregating Mutants, Irradiated Seeds. Burlington, N.C.: Carolina Biological Supply Company, Inc., 1973.

Name _____   Date _____

HOW ARE TRAITS TRANSMITTED IN CORN PLANTS?

You are to grow and observe corn seedlings which have a mutation that expresses itself phenotypically. The following will be determined from your observations:

- the mutation present (there may be more than one observable mutation in the seeds you are given);
- the phenotypic ratios of normal to mutant plants; and
- the mode of transmission of the mutant gene.

Safety Notes:

Keep all plant parts away from your mouth. Do not eat seeds; some may be treated with toxic pesticides or fungicides.

Wash your hands after each lab in which you handle seeds.

Setting Up Flats

1. Obtain experimental (mutant) and control (normal) seeds from your teacher. These are to be grown in separate flats.
2. Label two flats with your initials, the date, and the type of seed (experimental or control).
3. Mix equal parts of potting soil and vermiculite, and fill your flats with the mixture.
4. Plant 100 seeds of each type in separate flats, and cover the seeds with ½ inch of soil.
5. Moisten seeds with plant food mixture. "Feed" them every two or three days to keep them moist.
6. Germination will be evident within 4 to 7 days after planting. The mutation, however, will not be visible until 10 to 14 days after planting.

   NOTE: Information about germination and planting of mutant seeds © *Seedlings for Genetics: Segregating Mutants, Irradiated Seeds*. Burlington, N.C.: Carolina Biological Supply Company, Inc., 1973, used by permission.

7. Compare mutant and normal plants 14 days after planting. Record the information in Table 12–1, then answer the questions following the table.

Table 12–1

Type of Seedling	Type of Mutation	Number of Normal Plants	Number of Mutant Plants
Experimental			
Control			

Questions

1. Did all of your seeds germinate? If not, what percentage in each flat germinated? _____

_____

_____

HOW ARE TRAITS TRANSMITTED IN CORN PLANTS? (continued)

2. How do you know which of your experimental plants are phenotypically normal, if any, and which are mutants? _____

_____

_____

_____

_____

3. Describe the mutation(s) visible in your experimental seeds. _____

_____

_____

_____

_____

4. What is the phenotypic ratio of normal to mutant plants? _____

_____

5. Is it possible to determine from the data you have whether the gene(s) for the mutation(s) is (are) recessive or dominant? If so, how? If not, how could you determine this? _____

_____

_____

_____

_____

6. Using a Punnett square, show how this mutation is transmitted.

Name _____   Date _____

TRANSMISSION OF TRAITS EVALUATION QUESTIONS

1. In the fruit fly *Drosophila*, the gene for long wings (L) is dominant to the gene for dumpy wings (*l*). The gene for normal eye (N) is dominant to the gene for vermilion eye (n). A fruit fly having dumpy wings and normal eyes, known to be heterozygous for the eye trait, is crossed with a fly with vermilion eyes and long wings, known to be heterozygous for the wing trait. Write the percentages below of all the possible phenotypes of offspring produced when these two flies are mated.
   - _____ a. long wings, normal eyes
   - _____ b. long wings, vermilion eyes
   - _____ c. dumpy wings, normal eyes
   - _____ d. dumpy wings, vermilion eyes

2. In garden peas, tall vine is dominant and short vine is recessive. A homozygous tall plant is crossed with a short plant. The F1 plants are allowed to undergo self-fertilization. On the back of this sheet, show the predicted phenotypic and genotypic ratios of the F2 generation.

3. In a certain population of adults, the gene for laziness (L) is dominant to the gene for a studious attitude (*l*). The gene for the tendency to gossip (G) is dominant to the gene for minding one's own business (g). Two of the adults marry. The man is a lazy gossiper, known to be heterozygous for both traits. The woman is also heterozygous lazy, but minds her own business. Complete the diagrams below to show all the possible combinations of gametes produced by each individual. All of the chromosomes of the individual are not represented, only those which carry these traits. Assume that the traits are not linked.

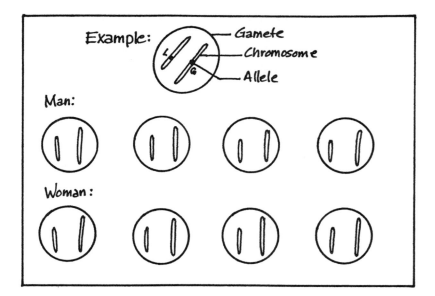

ACTIVITY 13

WHAT ARE SOME INTERACTIONS BETWEEN GENES AND THE ENVIRONMENT?

Goals

After completing this activity, students will:

1. better understand how changes in phenotype are produced by environmental factors; and

2. better understand the interplay between genes and environment in determining an organism's phenotype.

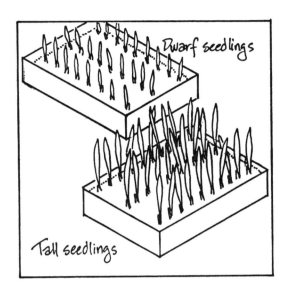

Synopsis

Students will design and carry out experiments to test the effects of various environmental factors on the phenotype and growth of mutant seeds. They will then speculate on the extent to which each of the effects was determined by genes, and to what extent by environment.

Relevant Topics

Mendelian genetics
DNA, RNA, and protein synthesis
mutation
environmental science
seed germination

Age/Ability Levels

grades 9–12, most ability levels

Entry Skills and Knowledge

Before participating in this activity, students should have completed the lab entitled "How Are Traits Transmitted in Corn Plants?".

Materials

photocopies of student worksheets: "What Are Some Interactions Between Genes and the Environment? Experiment Guidelines,"
"What Are Some Interactions Between Genes and the Environment? Planning Worksheet," and
"What Are Some Interactions Between Genes and the Environment? Data and Conclusions Worksheet."

per pair of students
four flats, at least 2 square feet in size
vermiculite
potting soil
200 mutant seeds
200 normal seeds
one marking pencil
two small trays for exposing seeds

per class
labeling tape
watering bottles
plant food
various types of household and laboratory chemicals
access to conventional and microwave ovens and a freezer
ultraviolet light
gibberellic acid solution
safety goggles
lab aprons

DIRECTIONS FOR TEACHERS

NOTE: If you are teaching younger or lower ability students, you may modify this lab by:
- using seeds with mutants involving monohybrid crosses only; and
- using seeds with only one visible mutation.

Expected Results

Results will vary, depending on the type of treatment and duration of exposure. The following treatments produce changes as described:

1. Ultraviolet light can induce mutations which may be evident as changes in rate of germination and appearance of plants.
2. Gibberellic acid breaks dormancy in some plants, speeding up germination.
3. Periods of long exposure to harsh chemicals may destroy areas of the seed, such as the aleurone layer, which produce enzymes necessary for plant growth. If this happens, changes in germination rate and appearance may be evident.

Preparation

1. Obtain different types of seeds with the same mutation. (The following are available from the agricultural department of a university or from a biology supply house, such as © Carolina Biological Supply Company; 2700 York Road, Burlington, N.C 27215, used by permission.)

    —green: albino (3:1); a monohybrid cross with a recessive mutant: corn, sorghum, and tobacco

    —green: yellow green: yellow (1:2:1); a blended inheritance cross: soybean and tobacco

    —green: albino (9:7); a dihybrid cross with two factors controlling albinism: corn; it would be interesting to compare this with the monohybrid albino cross.

    —tall: dwarf (3:1); a monohybrid cross with a recessive mutant: corn

    —dwarf (100%); a special mutant: corn; it would be interesting to compare this with the monohybrid dwarf plant.

2. Purchase normal seeds from a seed store or from a biological supply house.

3. Place mutant seeds in envelopes, 100 of each mutant per envelope. Label each envelope as to the type of seed and mutation it contains. For a given class, prepare envelopes containing different types of seeds with the same mutation to allow meaningful comparisons. For example, use both corn and sorghum, green: albino (3:1) seeds.

4. Place normal seeds in envelopes (100 per envelope). Label them as to seed type— "normal corn seeds," for example.

5. Obtain lab materials, including the various chemicals and equipment needed for student exposures of seeds.

6. Photocopy student guidelines worksheets, with extra copies of the data and conclusions worksheet for students who may wish to observe more than one plant characteristic.

7. Make up a 1000 ppm solution of gibberellic acid[1] as follows:

—dissolve 1 g of crystalline gibberellic acid in 5 ml of 95% ethanol;

—add a few drops of baby shampoo as a wetting agent; and

—bring the volume to 1 l with warm distilled water

SAFETY NOTES:

- If you use any seeds which are treated with pesticides or fungicides, wash them before using.
- Caution students not to eat seeds and to keep all plant parts out of their mouths. Have them wash their hands at the end of each lab period. This is especially important when handling sorghum and tobacco, since parts of these plants are toxic.
- Remind students to exercise caution in handling all laboratory chemicals. Do not ingest or inhale them directly and avoid contact with eyes, clothing, and skin. Acids and bases, in particular, can severely damage skin and eyes. Flood with water immediately if contact occurs.
- If laboratory spills occur, do the following:

—if a fire hazard exists, extinguish all flames immediately;

—dilute the spill with water;

—apply a commercial absorbent such as kitty litter to the spill; and

—clean the area thoroughly with soap and water, then mop it dry. Wear rubber gloves, and use a dustpan and brush while cleaning up.

- Have students wear goggles and lab aprons when handling laboratory chemicals.
- Keep alcohol away from open flames. Do not heat it directly.
- When heating seeds in a conventional oven, use hot pads when handling them. Wait until they are cool before planting them.
- When using ultraviolet lights, caution students not to look at them, since they can cause eye damage. Special glasses which protect the eyes from UV radiation can be purchased. UV lights can build up heat, so make sure to let them cool before putting them away.
- Check microwave ovens to be sure they do not emit dangerous levels of radiation. Caution students not to stand near a microwave oven while it is running.

---

[1] (Excerpt on gibberellic acid solution adapted from *A Sourcebook for the Biological Sciences*, third edition, by Evelyn Morholt and Paul S. Brandwein, copyright © 1986 by Harcourt Brace Jovanovich, Inc., reprinted by permission of the publisher.)

Teaching the Activity

1. Have students perform the experiment entitled "How Are Traits Transmitted in Corn Plants?" as a prerequisite to this lab.

2. Have students work in groups of four, sharing four flats, if space and supplies are limited. Otherwise, have students work in groups of two, sharing four flats.

3. Distribute seeds, giving each group two envelopes of the same type of mutant seeds (for example, green: albino corn 3:1), and two envelopes of normal seeds of that mutant (for example, normal corn).

4. Inform students that they may expose plants to another environmental agent not listed, provided it is available and has your approval.

5. Provide class time for students to design their experiments, giving assistance as needed. Mark your approval for numbers 3 and 9 on the planning worksheet. Be sure each group has included:

   —two control groups:

      normal seeds, unexposed
      mutant seeds, unexposed

   —two experimental groups:

      normal seeds, exposed
      mutant seeds, exposed

6. Students may expose seeds in class or at home.

7. Allow time in class for students to:

   ● set up flats on day 1;

   ● "feed" seeds every two or three days; and

   ● record data after mutations are visible. Since mutations will become visible on different days for the various types of seeds, be sure to plan some other activity for students who are not making observations.

8. Compare results as a class. When students have completed their worksheets, review the questions with them, calling for student responses.

Answers to Questions on the Student Planning Worksheet

1. through 5. Answers will vary; students are given suggestions on their worksheets for variables to test and characteristics to observe; see item 5 in the teacher's section ("Teaching the Activity") for suggested experimental and control groups.

6. A sample hypothesis and rationale might be: "When exposed to ultraviolet light for three minutes, mutant corn seeds will fail to germinate. Normal corn seeds exposed to UV light under the same conditions, however, will germinate. The reasoning for this is that mutant seeds are "abnormal," and less likely to be able to resist UV damage. Normal seeds, by contrast, can tolerate a certain amount of damage without it being lethal."

8. Answers will vary, but should be based on information in Table 13–1.

Answers to Questions on the Data and Conclusions Worksheet

1. See "expected results" section; results will vary.

2. a. through c. Results will vary.

3. Answers will vary.

4. Answers will vary.

5. Answers will vary.

6. Evidence for changes in genes include the following: a difference in the ratios of normal to mutant plants and changes in morphology of seedlings.

7. All changes are the results of an interaction between the two. For example:

   ● A change in the genes caused by UV light is the result of an environmental stimulus, but the change occurred in the genetic material. Also, plants differ in susceptibility to this type of mutation because of genetic differences.

   ● Destroying an essential part of the seed by a harsh chemical was caused by an environmental agent, but genes are responsible for necessary enzymes made by the destroyed tissue.

   ● If one mutant plant responds differently to the same treatment than another seed type with the same mutation, this is evidence that there is a difference in the genes of these two plants which affects the response. The changes, nonetheless, were induced by environmental agents.

8. See answer to 7.

Follow-up Activities

● Have students apply gibberellic acid to germinated seedlings at various stages of growth. Compare the results with application of the hormone to seeds which have not been germinated. Then:

   —compare the responses of the two groups and explain any differences; and

   —cite genetic and environmental factors which might be responsible for these differences.

● Have students work through the activity, "How Are Eukaryotic Genes Regulated?", proposing a molecular mechanism for genetic changes which occurred in this experiment with seeds.

● Have students repeat *this* experiment, exposing germinated seedlings to various environmental agents. Have them make comparisons across plant type and across different stages of growth.

Evaluation

Method of Evaluation	Goal(s) Tested
1. Completion of student worksheets and data charts	1,2
2. Participation in class discussion	1,2

References

1. *Seedlings for Genetics: Segregating Mutants, Irradiated Seeds.* Burlington, NC: Carolina Biological Supply Company, Inc., 1973.

2. Morholt, E., and Brandwein, P. F., *A Sourcebook for the Biological Sciences* (3rd ed.). New York: Harcourt Brace Jovanovich, 1986.

WHAT ARE SOME INTERACTIONS BETWEEN GENES AND THE ENVIRONMENT?
Experiment Guidelines

You have learned how certain mutant traits are transmitted in corn seeds. Now you are going to design and carry out an experiment in which you:

1. expose various types of mutant seeds to environmental factors;
2. plant the seeds and observe any changes in their growth or appearance; and
3. try to determine to what extent the changes were caused by factors in the environment and to what extent by genes.

SAFETY NOTES:

- Keep all plant parts away from your mouth. Do not eat seeds; some may be treated with toxic pesticides or fungicides. Parts of sorghum and tobacco are toxic.
- Wash your hands after each lab in which you handle seeds.
- Exercise caution in handling all laboratory chemicals. Do not ingest or inhale them directly, and avoid contact with eyes, clothing, and skin. Acids and bases, in particular, can severely damage skin and eyes so it is best to use weak acids and bases. Flood with water immediately if contact occurs.
- If laboratory spills occur, inform your teacher at once.
- Wear goggles and lab aprons when handling laboratory chemicals.
- Keep alcohol away from open flames. Do not heat it.
- When heating seeds in a conventional oven, use hot pads when handling seeds. Wait until they are cool to plant them.
- When using ultraviolet lights, do not look at them, since they can cause eye damage. UV lights can build up heat, so make sure to let them cool before putting them away.
- Do not stand near a microwave oven while it is in use.

WHAT ARE SOME INTERACTIONS BETWEEN GENES AND THE ENVIRONMENT
(continued)

Designing Your Experiment

You have control over these aspects of the experiment:

1. The environmental agent to which you expose seeds. Here are some suggestions:

    —heating in a conventional oven

    —microwaving

    —freezing

    —exposure to ultraviolet light

    —rinsing in a chemical such as a detergent, an acid, a base, acetone, or alcohol

    —exposing seeds to gibberellic acid, a plant hormone

2. The duration of exposure.
3. The types of changes you will monitor (you may look at more than one), such as:

    —whether germination occurs

    —plant height

    —changes in appearance of plants

    —time required for germination

    —changes in ratios of normal to mutant plants

These are your limitations:

1. Your experiment must include control seeds, as well as experimental ones.
2. You must use the seeds supplied by your teacher.
3. You are allowed 200 mutant seeds (100 per flat) and 200 normal ones.

Name _____   Date _____

WHAT ARE SOME INTERACTIONS BETWEEN GENES AND THE ENVIRONMENT?
PLANNING WORKSHEET

Indicate your choices for each of the following: (except mutant seeds, which your teacher will choose)

1. experimental (mutant) seeds: _____

   _____

2. control seeds: _____

   _____

3. variable: (Obtain teacher approval on this before proceeding.) _____

   _____

   _____

4. plant characteristic(s) you will observe: _____

   _____

5. expected results:

   a. experimental seeds: _____

      _____

   b. control seeds: _____

      _____

6. Write your hypothesis and a one-sentence rationale for it: _____

   _____

   _____

7. Transfer the information above to Data Table 13–2. (Use a separate table for each characteristic you observe.)

8. List below the steps you will follow in carrying out your experiment. Follow the same general procedure as in the previous lab regarding planting seeds, but refer to the information in Table 13–1 for specific instructions regarding your seeds.

WHAT ARE SOME INTERACTIONS BETWEEN GENES AND THE ENVIRONMENT?
PLANNING WORKSHEET (continued)

Table 13–1

seed type	# per sq ft	soil depth	time to germinate	time to see mutations
corn	50	½″	4–7 days	10–14 days
sorghum	100	¼″	3–6 days	7–10 days
tobacco	100	¼″	7–10 days	12–14 days
soybean	2″ apart, in rows 3″ apart	½″	7–10 days	12–14 days

(Information on germination and planting of seeds © *Seedlings for Genetics: Segregating Mutants, Irradiated Seeds*. Burlington, N.C.: Carolina Biological Supply Company, Inc., 1973, used by permission.)

Your Experimental Procedure:

_____

_____

_____

_____

9. Obtain teacher approval before going further.
10. Record your results on the chalkboard, then answer the questions which follow Table 13–2 on the Data and Conclusions worksheet.

Teacher Approval _____

Name _____ Date _____

WHAT ARE SOME INTERACTIONS BETWEEN GENES AND THE ENVIRONMENT? DATA AND CONCLUSIONS WORKSHEET

Fill out Table 13–2, then answer the questions. (Attach additional tables for additional observed characteristics.) You may expand the table to include "actual results" observations on each day, if you wish.

Table 13–2

Seed Type	Mutation	Treatment	Observed Characteristics	Expected Results	Actual Results
Experimentals 1					
2					
Controls 1					
2					

Questions

1. What changes, if any, did you observe in experimental as compared with control seeds as a result of the experimental treatment? _____

_____

_____

2. How did the experimental mutant seeds compare with:

a. the unexposed mutant ones? _____

_____

b. the unexposed normal ones? _____

_____

c. the exposed normal ones? _____

_____

WHAT ARE SOME INTERACTIONS BETWEEN GENES AND THE ENVIRONMENT?
(continued)

3. Did your results support or refute your hypothesis?

_____

4. How might you explain the results of your experiment?

_____

_____

_____

_____

5. How do your results compare with those of other class members—especially those who used a different seed type with the same mutation?

_____

_____

6. Do you have evidence to indicate that the genes of your seeds were changed as a result of the treatment? Explain.

_____

_____

7. Are any of the changes you observed strictly the result of the plant's genes? Are any strictly the result of the environment?

_____

_____

8. Explain the possible roles of genes and environment in determining these changes.

_____

_____

_____

_____

_____

_____

ACTIVITY 14 ———————————————
WHAT IS THE EFFECT OF RADIATION ON PLANTS?

Goals
After completing this activity, students will:

1. be able to identify the effects of a mutagenic agent on plant seedlings; and
2. better understand how mutations occur.

Synopsis
Students will predict the effects of varying dosages of radiation on the morphology and growth of plant seedlings, then test their predictions with experiments they design.

Relevant Topics
radiation
mutation
protein synthesis
DNA replication

Age/Ability Levels
grades 9–12, most ability levels

Entry Skills and Knowledge
Before participating in this activity, students should be able to:

1. explain DNA replication and protein synthesis;
2. define the term *mutation*; and
3. describe general concepts of Mendelian genetics.
   NOTE: These skills are not necessary to successful performance in this lab, but they will help make it a richer experience for students.

Materials
photocopies of student worksheets: "What is the Effect of Radiation on Plants?" and "Radiation's Effects Data Table"

per pair of students
six small pots for planting
one flat (in which to store the pots)
one marking pencil

per class
potting soil
vermiculite
plant food
watering bottles
trowels
labeling tape
irradiated seeds (see teacher preparation section for details)
control seeds

Directions for Teachers

Preparation

1. Obtain irradiated seeds. There are two basic routes for this:

   —Obtain seeds that have been irradiated from a commercial biological supply house. (Carolina Biological Supply Company carries barley, marigold, oat, peanut, radish, squash and tobacco seeds in exposures of from 20,000 rads to 50,000 rads of gamma radiation. Control seeds for each are also available; used by permission.)

   —Irradiate your own seeds using an X-ray machine in a dentist's or doctor's office.

2. Obtain student lab materials.

3. Photocopy student worksheets, with extra copies of the data table for students who may wish to observe more than one plant characteristic.

SAFETY NOTES:

- If you use any seeds which are treated with pesticides or fungicides, wash them before using.

- Caution students not to eat seeds and to keep all plant parts out of their mouths. Have them wash their hands at the end of each lab period. Parts of tobacco are toxic.

- If you expose seeds to alpha, beta, or gamma radiation using a source in a doctor's office or scientific laboratory, do so only under strict supervision of a trained technician. Stay well away from the source of radiation, in a shielded area, and do not handle the source. These types of radiation can damage human tissues.

Expected Results

Radishes are hardier than marigolds. You may see germination of radishes, but not other seeds, at highest dosages. (Expected results © *Seedlings for Genetics*: *Segregating Mutants, Irradiated Seeds*, Burlington, N.C.: Carolina Biological Supply Company, Inc., 1973, used by permission.)

Teaching the Activity

NOTE: This lab can be used with lower ability students, with no reference to mutation or DNA. If you choose this approach, omit the last three discussion questions under item 8 below.

1. Briefly introduce the lab and distribute student worksheets. Define radiation and ask students what effects it might have on plants.

2. Provide students with a list of the seeds available, and the dosages of radiation to which they have been exposed.

3. Have students work in pairs. Allow time for them to design their experiments and set up data tables.

4. Provide help as needed, but have students develop their own hypotheses.

5. When students are ready, show them how to prepare, label, and "feed" seeds. Store them in flats in a lighted area.

6. Allow some time in class each day for making and recording observations.

7. At least once each week, ask students for informal oral progress reports.

8. Compare results as a class. Below are some questions to stimulate discussion. Answers are given in parentheses.
   - What levels of radiation seem to be more harmful to seeds—high or low? (high)
   - Did the various types of seeds respond differently to the radiation? If so, how might this be explained? (See "expected results"; different genes may be affected in different ways by the radiation in various plants.)
   - Are all the changes you observed "bad" for the plant? Were any of them "good"? How do you define "good" and "bad" from the point of view of the plant? (A "good" change is one that helps the plant survive; students may see taller than normal plants, conceivably a "good" change, indicating a possible change in one or more genes regulating growth.)
   - Are the changes you saw permanent? How could you find out? (They are all permanent for the plant being observed. To determine whether they can be passed on to offspring, students would need to breed the plants with other irradiated ones exhibiting similar changes.)
   - How do you think the radiation caused these changes? (If students have not already learned about mutation, this is an excellent time to introduce it. Several kinds of mutation will be represented here—point mutations and gross chromosomal abnormalities.)
   - Might there be changes in the DNA of these seedlings, resulting from the radiation, which did not show up in these plants? How could you find out? (Yes; some mutations are recessive. Breeding plants for several generations would reveal these traits in offspring.)

9. If possible, have students perform the experiment described in the first follow-up activity.

Answers to Questions on Student Worksheets

Designing Your Experiment

1. through 4. answers will vary; suggestions are given on the first student worksheet for variables and plant characteristics to be observed

5. answers will vary for expected results; a sample hypothesis and rationale might be: Barley seeds exposed to 20,000 rads of radiation will germinate; those exposed to 50,000 rads will not. The rationale for this is that more damage is likely to be done to a plant receiving 50,000 rads as compared to 20,000 rads, making it less likely that the 50,000 rads plant will germinate.

Questions

1. through 3. Answers will vary.

Follow-up Activities

- Have students transfer small seedlings from this experiment to individual large pots, or to a garden outdoors, and observe them for a longer period of time to follow radiation-induced changes.
- Have students work through the exercise entitled "How Do Mutations Occur?".
- Have students breed plants from this experiment to determine whether the changes resulting from radiation are passed on to offspring.

- Have students research ways an organism might "correct" damage incurred through mutation. Begin by studying DNA repair enzymes.
- Have students read about xeroderma pigmentosum, a human skin disease caused by a lack of a DNA enzyme which repairs mutations.
- Irradiate plants at different stages of growth, and compare their appearance with that of untreated plants.
- In 1983, NASA sent tomato seeds into orbit around the earth, with the assistance of astronauts of the Space Shuttle. These seeds have been retrieved, and are being distributed to American teachers as long as supplies last. For information on the seeds, write: SEEDS Project; National Aeronautics and Space Administration; Code XEO; Washington, DC 20546. Have students repeat the experiment described here using these seeds.
- If it is possible to irradiate your own seeds, repeat this experiment, comparing the effects of:

  —different types of radiation (alpha, beta, gamma, X-rays, ultraviolet) on seeds; and

  —different dosages of exposure of each type of radiation on seeds. Follow all safety precautions for exposure to UV and alpha, beta, and gamma radiation, as detailed earlier in this section.

- Have students do library research on the effects of nuclear bombs on living organisms. *NOTE*: This topic will both intrigue and disturb students, so be sure to discuss their feelings about nuclear war with them.

Evaluation

Method of Evaluation	*Goal(s) Tested*
1. Completion of student worksheet and data chart	1, 2
2. Participation in class discussion	1, 2
3. Question 1	2

Question 1

Given the following strand of DNA nitrogen bases, indicate the changes that would occur in the mRNA and tRNA sequences if the second nucleotide on the left and the last two nucleotides on the right were deleted from the DNA through point mutations:

DNA: T   A   C   G   A   C   A   G   G   C   T   C
mRNA: __ __ __ __ __ __ __ __ __
tRNA: __ __ __ __ __ __ __ __ __

Answers

DNA: TCGACAGGC
mRNA: AGCUGUCCG
tRNA: UCGACAGGC

Reference

Seedlings for Genetics: *Segregating Mutants, Irradiated Seeds*. Burlington, N.C.: Carolina Biological Supply Company, Inc., 1973.

Name _____ Date _____

WHAT IS THE EFFECT OF RADIATION ON PLANTS?

You will design an experiment to determine the effects of radiation on different types of seeds.

These are the aspects of the experiment over which you have control:

1. the type of seed and dosage of radiation. Your teacher will provide you with a list of available seeds and dosages.
   For example, you might:
   - compare the effects of different dosages of radiation on a single type of plant;
   - compare the effects of one dosage of radiation on different types of plants; OR
   - look at a combination of these two.

2. the characteristic(s) of the plant to be observed. For example, you might look at:
   - numbers of plants that germinate;
   - plant height;
   - types of plant abnormalities; and
   - numbers of abnormal plants.

Your limitations are that:

1. You must include control seeds, in addition to experimental ones;
2. each lab group is allowed six pots; and
3. you should plant no more than six seeds in each pot.

Safety Notes:

Keep all plant parts away from your mouth. Do not eat seeds; some may be treated with toxic pesticides or fungicides. Parts of tobacco are toxic.

Wash your hands after each lab in which you handle seeds.

Designing Your Experiment

To design your experiment, indicate your choices for each of the following:

1. variable(s): _____

2. experimental plants:

   type of plant   *dosage of radiation*

   a. _____   _____

   b. _____   _____

   c. _____   _____

3. control plants:

   type of plant   *dosage of radiation*

   a. _____   _____

   b. _____   _____

   c. _____   _____

4. plant characteristic(s) to be observed:

   a. _____

   b. _____

   c. _____

5. expected results: _____

_____

Write a hypothesis indicating the anticipated outcome of your experiment. Then, write a one sentence rationale for it.

_____

_____

Data Setup

Glue together all three pages of the Data Table, and record the information above on it. Use a separate chart for each plant characteristic you observe.

Procedure

1. Label pots, indicating plant type, dosage, the date, and your initials.
2. Fill pots with a mixture of potting soil and vermiculite, in equal parts.
3. Plant seeds ½ inch deep, covering them with soil.
4. Moisten plants with plant food mixture. They should be "fed" about every third day.
5. Store plants in a lighted area.
6. Observe plants every day for three weeks, if possible. Record observations in the Data Table, then answer the questions that follow.

Questions

1. How did your actual results compare with your expected results? _____

_____

_____

2. Was your hypothesis supported or refuted? Explain. _____

_____

3. How might your results be explained? _____

_____

_____

Name _____ Date _____

RADIATION'S EFFECTS—DATA TABLE

Type of Plant	Dosage	Plant Characteristic	Expected Results	Actual Results: DAY 1
Experimentals				
1.				
2.				
3.				
4.				
5.				
Controls				
1.				
2.				
3.				

Attach here

DATA TABLE, (continued)

2	3	4	5	6	7	8	9	10	11

Attach here

DATA TABLE, (continued)

12	13	14	15	16	17	18	19	20	21

ACTIVITY 15

HOW DO MUTATIONS OCCUR?

Goals

After completing this activity, students will:

1. know the three types of point mutations occurring in DNA;
2. better understand the process of protein synthesis; and
3. understand the mechanism by which mutations produce changes in the cell.

Synopsis

Students will discover the three types of point mutations (deletion, addition, and substitution) by simulating these using bits of colored paper.

Relevant Topics

genetics
evolution

Age/Ability Levels

grades 9–12, average to gifted

Entry Skills and Knowledge

Before participating in this activity, students should be able to:

1. explain the significance of DNA in controlling cellular activities;
2. cite the number of DNA bases required to code for an amino acid;
3. describe the relationship between amino acids and proteins;
4. describe briefly the process of protein synthesis; and
5. list some examples of proteins.

Materials

student worksheets: "How Do Mutations Occur?"
envelopes (two per pair of students)
colored construction paper; four different colors are needed
(optional) a commercial model which demonstrates protein synthesis

DIRECTIONS FOR TEACHERS

Preparation

1. Make (or have students make) the DNA and RNA "bases" out of colored construction paper by following these guidelines:
   a. Cut each base to a size of ½ inch square.
   b. Color code bases as follows:

131

DNA BASES	RNA BASES	COLOR
adenine	adenine	blue
guanine	guanine	green
cytosine	cytosine	red
thymine	uracil	yellow

   c. Label each base on both sides in dark ink, *A* for adenine, and so forth.

   d. Make at least ten duplicates of each DNA and RNA base for each envelope. This allows extra bases to be added or substituted in the "mutations."

2. Place DNA and RNA bases in separately labelled envelopes.

3. Photocopy student directions and codon and anticodon tables.

4. Be sure students have mastered entry skills. Use a commercial model illustrating protein synthesis, if one is available.

Teaching the Exercise

1. Review mutation briefly with students, defining it as a change in an organism which can be passed on to its offspring.

2. Distribute student directions and codon and anticodon tables. Be sure each group of students receives one envelope each of DNA and RNA bases. Briefly explain the assignment.

3. Have students work in pairs. Provide help as needed, asking such questions as these:

   • In what ways can you change the order of the DNA bases?

   • How could you change the sequence by changing the number of DNA bases in front of you?

4. A correct simulation is illustrated as follows:

Assume bases are arranged in the sequence shown in Figure 15–1.

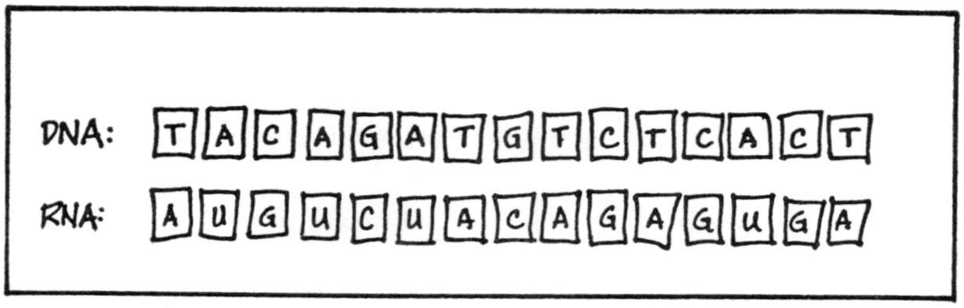

DNA: T A C A G A T G T C T C A C T
RNA: A U G U C U A C A G A G U G A

FIGURE 15–1

If a deletion of the fourth nucleotide base from the left in the DNA section above were to occur, all bases would be ordered again to create a readable sequence, simulating the repair process of the cell. The new sequence would then read as shown in Figure 15–2.

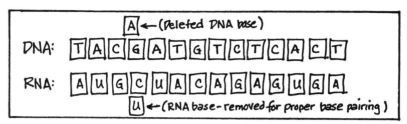

FIGURE 15-2

When bases are added or deleted, the result is a one- or two-base codon at the end of the sequence—not enough to specify an amino acid. Instruct students to ignore these "extra" bases.

5. Call for volunteers to present the three types of mutations to the class. Ask students to explain why a change in DNA is "permanent" for the cell, while a change in RNA is only temporary. Mutations are defined as changes in DNA only.

6. Have students return bases to the proper envelopes, then collect the bases and worksheets.

Answers to Questions on Student Worksheet

1. Answers will vary.
2. Messenger RNA; answers will vary for the RNA sequence.
3. a. deletion
   b. addition
   c. substitution
4. Answers will vary.
5. Probably not, but answers will vary.
6. a and b. The protein normally specified by that sequence would not be made.

   c. The peptide would terminate at the stop codon.
7. a and b. Enzymes, muscle protein, cell membrane protein, antibody, hemoglobin; a mutation could cause a change in shape and function in any of these proteins, potentially preventing the cell from carrying on its normal processes.

Follow-up Activities:

- Have students count the number of hydrogen bonds between the DNA and RNA base sequences on their worksheets.
- Have students read about genetic abnormalities other than those covered in this exercise. Examples include the following:

  —trisomy (a cell with one too many chromosomes)

  —monosomy (a cell with one too few chromosomes)

  —chromosome translocation (a cell with all or part of one chromosome attached to another)
- Invite speakers such as genetic counselors, doctors, and nurses in to discuss genetic diseases caused by chromosome anomalies, such as those listed below:

  —Klinefelter's syndrome

  —Turner's syndrome

  —Down's syndrome

- Have students research the effects of certain drugs and harmful chemicals—such as DES and LSD—on mammalian chromosomes.
- Have advanced students explain how a point mutation can cause a profound change in the organism. For example, sickle cell anemia is caused by a change in the base sequence specifying one amino acid of the hemoglobin molecule.
- Have students read about the teratogenic effects of thalidomide, a sedative given to pregnant women in the early 1960s. It was withdrawn from the market because it caused fetal abnormalities. Have students research the reasons why thalidomide was never marketed in the United States, then decide whether they think it is wise to obtain FDA approval before marketing drugs, even though testing requires time.
- Certain environmental agents, such as preservatives in food, are known to cause mutations. Have students research some of these and their possible mechanisms of action.

Evaluation

Method of Evaluation	Goal(s) Tested
1. Completion of student worksheets.	1,2,3
2. Question 1	1,2
3. Question 2	1,2
4. Question 3	3

Question 1

Write a short DNA base sequence on the chalkboard. Using their codon and anticodon tables, have students:
a. write the correct messenger RNA sequence complementary to it; and
b. write the correct amino acid sequence specified.

Question 2

Using the DNA sequence from question 1, have students write the resulting mRNA and amino acid sequences when:
a. the third nucleotide from the left is deleted;
b. a guanine is added immediately after the fifth nucleotide from the left; and
c. the first adenine appearing in the sequence has a guanine substituted for it.

Question 3

Have students list possible changes occurring in a given cell as a result of point mutations.

HOW DO MUTATIONS OCCUR?

Point mutations are those involving only one or a few bases in DNA. Using the nucleotide "bases" in the envelopes you are given, you will:

1. discover what the three types of point mutations are, and
2. simulate these mutations.

Simulating the Mutations

Complete the information below, using the bases in the envelopes provided by your teacher and the codon and anticodon tables, Tables 15–1 and 15–2.

TABLE 15–1. DNA Codon Table

second base

		A	G	T	C	
first base	A	AAA phe AAG phe AAT leu AAC leu	AGA ser AGG ser AGT ser AGC ser	ATA tyr ATG tyr ATT STOP ATC STOP	ACA cys ACG cys ACT STOP ACC trp	A G T C
	G	GAA leu GAG leu GAT leu GAC leu	GGA pro GGG pro GGT pro GGC pro	GTA his GTG his GTT gln GTC gln	GCA arg GCG arg GCT arg GCC arg	A G T C
	T	TAA ile TAG ile TAT ile TAC met	TGA thr TGG thr TGT thr TGC thr	TTA asn TTG asn TTT lys TTC lys	TCA ser TCG ser TCT arg TCC arg	A G T C
	C	CAA val CAG val CAT val CAC val	CGA ala CGG ala CGT ala CGC ala	CTA asp CTG asp CTT glu CTC glu	CCA gly CCG gly CCT gly CCC gly	A G T C

third base

amino acid abbreviations:

ala	alanine	leu	leucine
arg	arginine	lys	lysine
asn	asparagine	met	methionine
asp	aspartic acid (aspartate)	phe	phenylalanine
cys	cysteine	pro	proline
gln	glutamine	ser	serine
glu	glutamic acid (glutamate)	thr	threonine
gly	glycine	trp	tryptophan
his	histidine	tyr	tyrosine
ile	isoleucine	val	valine

HOW DO MUTATIONS OCCUR? (continued)

TABLE 15–2. RNA Anticodon Table

second base

		U	C	A	G	
first base	U	UUU phe UUC phe UUA leu UUG leu	UCU ser UCC ser UCA ser UCG ser	UAU tyr UAC tyr UAA STOP UAG STOP	UGU cys UGC cys UGA STOP UGG trp	U C A G
	C	CUU leu CUC leu CUA leu CUG leu	CCU pro CCC pro CCA pro CCG pro	CAU his CAC his CAA gln CAG gln	CGU arg CGC arg CGA arg CGG arg	U C A G
	A	AUU ile AUC ile AUA ile AUG met	ACU thr ACC thr ACA thr ACG thr	AAU asn AAC asn AAA lys AAG lys	AGU ser AGC ser AGA arg AGG arg	U C A G
	G	GUU val GUC val GUA val GUG val	GCU ala GCC ala GCA ala GCG ala	GAU asp GAC asp GAA glu GAG glu	GGU gly GGC gly GGA gly GGG gly	U C A G

third base

1. Using your codon table as a guide, arrange the DNA bases from the envelope into a sequence specifying five amino acids. Your only limitations are that:
   - met (methionine) must be the first amino acid; and
   - there must be a STOP codon at the end of the sequence.

   Write your DNA sequence:

   _____

   Write the amino acid sequence specified by this DNA:

   _____

2. Arrange the RNA bases opposite the DNA bases using your anticodon table as a guide. What type of RNA have you formed (messenger, transfer, or ribosomal)? _____.

   Write the RNA sequence:

   _____

3. Three "mistakes," (mutations), occur in the sequence of DNA bases which change both the messenger RNA and the protein sequence it specifies. Working with the DNA bases in front of you, rearrange them until you figure out how these errors occur. Write the three "mistakes" below:

   a. _____

   b. _____

   c. _____

HOW DO MUTATIONS OCCUR? (continued)

4. Make each of these "mistakes" one at a time using your DNA bases. After each mutation, return all bases to the original sequence before proceeding to the next one. Each time, arrange the RNA bases opposite the "mutated" DNA bases. In the spaces below, record the types of mutations, the DNA and RNA sequences, and the amino acid sequences:

MUTATION a: _____

DNA SEQUENCE: _____

RNA SEQUENCE: _____

AMINO ACID SEQUENCE: _____

MUTATION b: _____

DNA SEQUENCE: _____

RNA SEQUENCE: _____

AMINO ACID SEQUENCE: _____

MUTATION c: _____

DNA SEQUENCE: _____

RNA SEQUENCE: _____

AMINO ACID SEQUENCE: _____

5. Are the amino acid sequences the same in each of the three cases above? _____

6. What would be the result if a mutation:
   a. produced a STOP codon at the beginning of a DNA base sequence?

   _____

   b. omitted a START codon at the beginning of a DNA base sequence?

   _____

   c. inserted a STOP codon in the middle of a DNA base sequence?

   _____

7. List two examples of proteins:
   a. _____
   b. _____
   Explain how a mutation in the amino acids of one of these proteins might affect the cell.

   _____

   _____

   _____

   _____

section V

Health Store Products—Do They Work? You Find Out!

Health store product advertisers regularly bombard consumers with claims of the miraculous benefits of their over-the-counter products. Many substances sold by health stores are valuable when used appropriately. Some, however, have no benefit, and others are potentially quite harmful. For example, certain products that may be important dietary supplements in small doses, such as vitamin C, may be said to cure illnesses in potentially harmful megadoses. Likewise, fat soluble vitamins (A, D, E, and K) can be toxic when consumed in large amounts.

This project is investigative in nature, and is designed to allow students to use their knowledge of science to determine the validity of claims of health product advertisers.

Since, in many cases, the effectiveness of products in health stores has not been documented, at least not by some of the techniques included here, student investigations will very often constitute original research. Students are intrigued by this possibility. Interest in this activity will usually be high, since many students or their family members will be using one or more of these products.

How to Use This Project

If students are to pursue the complete investigation of products outlined in this section, the project will require 3 to 4 weeks of class time. Since many teachers will have difficulty setting aside such a large block of time from a biology course, below are suggestions for integrating it into a typical high school biology curriculum:

1. After students have studied nutrition, biological chemistry, or both, substitute this project for labs and other assignments normally associated with these subjects.
2. Substitute the project for part of a unit on human anatomy and physiology.
3. Introduce this project into your program as a new unit called "Consumer Biology." The time needed for it can be taken from other units on nutrition, biological chemistry, and human anatomy and physiology, since it represents an alternative vehicle through which students can learn about these subjects.
4. The "Health Store Project" provides excellent material for science fairs. If your curriculum dictates that students devote a certain amount of class time to science fair projects, have them spend this time on the "Health Store Project."

If, despite these suggestions, you are unable to work the project into your schedule, each of the labs described in this section is self-contained. The individual labs can be used to supplement units on nutrition, biological chemistry, and anatomy and physiology. Each requires that students form hypotheses, then design experiments to test them.

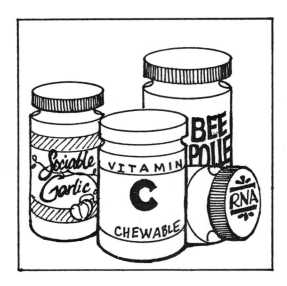

HEALTH STORE PRODUCTS—DO THEY WORK? YOU FIND OUT!
TEACHER OVERVIEW

This is a general overview of the whole project; you'll find additional information with each individual lab.

Goals

See individual experiments.

Synopsis

Students will test for themselves the validity of claims of products—such as dietary supplements and remedies for diseases—sold at "health stores." Investigations will include library research and student-designed experiments.

Relevant Topics

biological chemistry
nutrition
anatomy
physiology

Age/Ability Levels

grades 9–12, average to gifted

Entry Skills and Knowledge

Refer to individual labs.

Materials

student worksheet: "Student Overview"
Refer to individual labs for additional materials.

DIRECTIONS FOR TEACHERS

Background

As described in the introduction, this project is designed to enable students to test for themselves the validity of claims of products—such as dietary supplements and remedies for diseases—sold at "health stores." It has three components which should span a three to four week period. These include:

a. a laboratory students help design;
b. a written report; and
c. an oral class presentation of the findings.

   NOTES:

1. If time is lacking, omit one or more of the laboratories.

2. Simplify the project, if necessary, by omitting some of the questions on the written assignment.

Sequence of Events—Overview

Since this project is more complex than other activities in this book, the following overview of the sequence of activities is included, along with a suggested time frame, to help you plan. More explicit details on teaching the project are included separately for each lab.

1. Obtain products, information, and references several weeks in advance.

2. Obtain and prepare lab materials several days in advance.

3. Distribute information to students, and allow time for them to choose their products. Count this as day 1 of the student project.

4. Have students begin doing library research during class and/or on their own time (days 2–14).

5. Each of the experiments begins with a "practice" lab, then leads students through the design of their own experiments. The exception is the final set of labs. These are simple chemical tests and do not lend themselves readily to student designs. Have all students perform the "practice" labs as a class in the order listed so that they will have a basis for choosing how to test their products (days 3–11).

6. Allow students to choose the student-designed experiments or chemical tests they will perform with their products, and work on these individually in class (days 12–14). Since students should have mastered techniques in the practice labs, little supervision on your part should be required at this step.

7. Allow time for students to write up their results (days 15–19).

8. Have students present oral reports to the class, and collect written ones (days 20–23).

Preparation

1. Obtain products, information, and references:
   a. Obtain advertisements and pamphlets describing "health products" from stores selling them or from nutrition/health magazines on newsstands. These are valuable in determining the claims made about the product.
   b. Buy several of the advertised products.
   c. If possible, place references from the "Written Report Reference List" on reserve in your library. You may find additional valuable references not on this list in your library.

     d. Photocopy product advertisements, reference list, and student directions.

  2. Refer to each laboratory for information on preparation for individual lab experiments.

Written Report Reference List

1. Ames, B. N., "Dietary Carcinogens and Anticarcinogens," *Science*, 221, (1983), 1256–1262.
2. Gormley, T. R., Downey, G., & O'Beirne, D. *Food, Health and the Consumer*. New York, NY: Elsevier Applied Science, 1987.
3. Gussow, J. D., & Thomas, P. R. *The Nutrition Debate*: *Sorting Out Some Answers*. Palo Alto, CA: Bull Publishing Co., 1986.
4. Lewis, C., *Nutrition*: *Proteins, Carbohydrates, Lipids*. Philadelphia, PA: F. A. Davis Company, 1978.
5. McGee, H., *On Food and Cooking*. New York, NY: Charles Scribner's Sons, 1984.
6. Neuberger, A., & Jukes, T. H. (Eds.), *Human Nutrition*: *Current Issues and Controversies*. Englewood, NJ: Jack K. Burgess, Inc., 1982.
7. Pearle, L., *Junk Food, Fast Food, Health Food*: *What America Eats and Why*. Boston, MA: Clarion Books, 1980.
8. Peevey, L., & Smith, U., *Food, Nutrition, and You*. New York, NY: Charles Scribner's Sons, 1982.

Teaching the Project

(Additional teaching instructions accompany each lab activity.)

1. Distribute "Overview" worksheet, along with advertisements and reference list.
2. Provide time for students to read the information, then answer questions they may have.
3. Assign due dates for individual components of the project.
4. Have students begin doing library research on their own, during class, or both. Answers will vary on student written report questions.
5. Begin performing "practice labs" for each of the experiments in class. Explain to students that they should not decide which of the experiments they will use to test their products until they have been through all the practice labs. If you choose, you may allow advanced students to design and carry out an experiment not suggested here, with your approval. Directions for teaching each of the labs are given on the following pages.
6. Allow time for students to work individually, designing and performing their own labs.
7. Have students present their findings to the class orally.
8. After each presentation, briefly discuss student reactions.
9. Ask students to summarize what they learned from their research by answering this question: Would you use this product? Why or why not?

General Safety Notes:

- As with all laboratories, caution students to exercise care in handling all laboratory chemicals. They should not ingest or inhale them directly, and should avoid chemical contact with eyes and skin. Flood with water immediately if contact occurs.

- Warn students of specific safety precautions regarding iodine and ethanol.
- Check your local and county ordinances regarding disposal of chemicals. Have students dispose of them in the laboratory in separate glass containers.
- If laboratory spills occur, do the following:

  —If a fire hazard exists, extinguish all flames immediately;
  —dilute the spill with water;
  —apply a commercial absorbent such as kitty litter to the spill; and
  —clean the area thoroughly with soap and water, then mop it dry. Wear rubber gloves, and use a dustpan and brush while cleaning up.

- Have students wear laboratory aprons at all times and goggles whenever liquids are being heated.
- Allow plenty of time for students to clean up at the end of each lab period to reduce the chances of laboratory accidents. Have them clean their work areas thoroughly at the end of each lab.

Follow-up Activities

- Invite a nutrition expert in as a speaker to discuss the effects of certain "health" products on the body.
- Have students repeat this assignment for another product.
- Have students perform laboratory tests in addition to those suggested here to determine the effects of products. Some suggestions: 1) Testing the effect of the substance on planaria regeneration; or 2) Testing the effect of the substance on plant germination and/or growth.
- Have students repeat the laboratory experiments performed here with each individual product ingredient, comparing the effects of each ingredient separately and of the ingredients when mixed in various combinations.
- Consider having students mail letters summarizing results to the manufacturers of the products analyzed.

Evaluation (for entire unit)

Method of Evaluation	Goal(s) Tested
1. Completion of written report	1,2,3
2. Completion of student-designed laboratory	1,2,3
3. Presentation of written report	1,2,3
4. Participation in class discussion	1,2,3

References

1. Morholt, E., and Brandwein, P. F., *A Sourcebook for the Biological Sciences* (3rd ed.). New York, N.Y.: Harcourt, Brace, Jovanovich, 1986.

2. Wilbur, K. M., and McMahan, E. A., "Low Temperature Studies on the Isolated Heart of the Beetle, *Popilius disjunctus* (Illiger)," *Annals of the Entomological Society of America*, 51, no. 1, 1958.

STUDENT OVERVIEW: HEALTH STORE PRODUCTS—DO THEY WORK? YOU FIND OUT!

You will test the validity of the claims of one of the advertised "health products" on the information sheets provided by your teacher by researching the product in the library and in the laboratory. A list of reference materials will be provided by your teacher. You may use others in addition to these.

Assignment Requirements

I. WRITTEN REPORT summarizing your library and laboratory research: The questions below should be answered in your report for at least three product ingredients (either different ingredients of the *same* product, or separate products, each with only one ingredient). You may begin each section with the question it answers, underlining the question.

Questions:

1. What does your product contain? List ALL ingredients. Categorize each substance as either a vitamin, mineral, protein, fat, or carbohydrate.

2. What claims are made by the advertisers of the product?

3. How are the three ingredients you researched taken into the human body? Follow the pathway from its breakdown in the digestive system through its absorption in the small intestine, through the bloodstream, and into the cells, if possible.

4. What is the effect of each of the three ingredients on the human body? Include in your response answers to the questions below:

   • What are the benefits, if any?

   • What are the potential hazards, if any?

   • Are there differences in how it affects the body, depending on dosage? If so, describe these.

5. How is each of the three ingredients used in the body? For example, what type of molecule is it used to build—protein, vitamin, fat, or carbohydrate?

6. Summarize your hypothesis, laboratory experiment(s), and experimental data. Read the following pages before doing this.

7. Based on your answers to the questions above, are the claims made by the advertisers valid? Explain.

8. What effects do the following factors—time, temperature, light, and moisture—have on your product, if any? How might this influence how the product is stored and used?

9. List at least three references.

II. LABORATORY EXPERIMENT on your product: Perform ONE of the following experiments on your product. In the first two laboratories, you will be writing part of the experimental design yourself. The experiment and data should be summarized in the written report.

Beetle Heart Experiment

Test the effects of the product on the heart of the beetle, *Passalus cornutus*.

Vitamin C Determination   (appropriate only for products containing vitamin C)

Determine the vitamin C content of the product. (You may also test the product before and after treatment with such variables as temperature change, exposure to light, microwaves, and moisture.)

Chemical Tests for Biological Molecules

Perform the "biological molecules" tests on the product to determine whether it contains what is claimed. What other substances are present, if any, which are not listed on the package? (Separate worksheets detail individual lab procedures.)

III. ORAL PRESENTATION summarizing your findings: This should take no more than 5 minutes.

Due Dates:

In the spaces below, fill in the dates your teacher gives you indicating when each section of the project is due:

Due date 1: _____ choice of product and ingredients

Due date 2: _____ written report

Due date 3: _____ oral reports to begin

GENERAL SAFETY NOTE: Follow all safety precautions as directed by your teacher. Some of these are listed below:

- Keep all laboratory chemicals away from skin, clothing, and eyes; flood with water immediately if contact occurs.
- Do not ingest or directly inhale chemicals.
- If spills occur in the lab, inform your teacher at once and follow his or her directions for cleanup.

ACTIVITY 16 ———————————————————

WHAT ARE THE EFFECTS OF SUBSTANCES ON THE HEART OF THE BEETLE, *PASSALUS CORNUTUS*?

Goals

This activity will enable students to:

1. test the claims of "health store" products;
2. be more familiar with the ingredients, modes of action, and potential benefits and dangers of some of these substances;
3. make more informed choices about whether to use these products; and
4. become more adept at dissection of invertebrates

Note: For synopsis, relevant topics, and age/ability levels, see the general overview on page 140.

Entry Skills and Knowledge

Before participating in this activity, students should be able to:

1. describe briefly the process of human blood circulation;
2. measure liquids safely in the laboratory;
3. use a dissecting microscope; and
4. list some biological molecules and their functions.

Materials

student worksheets: "What Are the Effects of Substances on the Heart of the Beetle, *Passalus cornutus*? Practice Experiment" and
"What Are the Effects of Substances on the Heart of the Beetle, *Passalus cornutus*? Student-Designed Experiment"

for the entire class:
plastic shoe boxes with holes in lids
decaying deciduous wood
moist paper towels
timers or clocks with second hands
Ringer's solution
for each group of two students:
one beetle (*Passalus cornutus*, or Bessbug)
one dissecting microscope
one Petri dish
one dental-style wax plate (or wax-filled glass dish)
eight dissecting pins
one dissecting pan (or paper towels)
one pair of heavy scissors
one glass killing jar with lid containing: cotton, ethyl acetate
one Pasteur pipette with bulb

DIRECTIONS FOR TEACHERS

Background:

Since beetles are invertebrates with an open circulatory system, information obtained about their heart tissue must be applied to higher animals with caution. Beetles are, nonetheless, an excellent choice because: they are easy to maintain, the heart is clearly visible and simple to dissect, and similar experiments on vertebrates would be difficult for many students and parents to accept. National science education organizations and many science competitions now advise teachers not to allow students to use live vertebrates as test subjects.

Since the invertebrate heart is similar in some general ways to that of the vertebrate heart, it is possible to extrapolate to higher animals some results from an experiment on the beetle heart.

Preparation

1. Make up insect Ringer's solution by dissolving the following ingredients in distilled water, final volume of 1 l. Store in a refrigerator when not in use; allow it to warm to room temperature before using.

Compound	Amount (g/l)
NaCl	10.93
KCl	1.57
CaCl	0.85
MgCl2	0.17
NaHCO3	0.17

    (final pH should be 7.2)

2. Dissolve solid test substances for practice experiments in Ringer's solution. Suggested test substances include:
    a. substances which increase heart rate: stimulants such as NoDoz® and Vivarin,® and tea, soft drinks, coffee with caffeine (these can be compared to decaffeinated versions);
    b. substances which decrease heart rate: sleeping aids such as Sleep-eze,® Sominex,® and valerian root tablets.

3. Obtain beetles from a biological supply house, or find your own in rotting hardwood. The beetles ("bessbugs" or *Passalus cornutus*—previously known as *Popilius disjunctus*) are quite simple to keep alive for several weeks. Follow these steps:
    a. House beetles in a closed container such as a plastic shoe box with holes in the lid. Holes may be made by heating an ice pick and forcing it through the plastic. Keep the lid taped shut to prevent beetles from escaping. *Safety Note:* Exercise caution when heating the ice pick. Wear safety goggles, and keep the ice pick pointed away from you.
    b. Each shoe box should contain:

    • no more than seven beetles, fewer if possible;

    • decaying deciduous wood, such as oak; some will be shipped with the beetles if they are purchased; more will need to be added if they are kept for several weeks; and

● moist paper towels partially covering the wood.

c. Add new moist paper towels every 2 to 3 days. Add more wood as needed. Beetles consume the wood, and an accumulation of "sawdust" that their digestive systems have processed will soon be visible.

d. Set up killing jars with absorbent cotton soaked with ethyl acetate.

Note: As an alternative to beetle dissection, some teachers may prefer to use *Daphnia* (water fleas). The primary advantage of *Daphnia* is that it offers a way of assessing the product effect on an intact living system, which is more realistic than the dissected beetle heart. Disadvantages are that it does not offer students practice in dissection, and *Daphnia* hearts are smaller and more difficult to see than the beetle hearts. If you choose to use *Daphnia*, Morholt and Brandwein (1986) suggest immobilizing them by placing them in the well of a depression slide which has had petroleum jelly applied to it. Aquarium water can then be added, and the heart rate measured as it is done with beetle hearts. Ringer's solution will not be needed. For more information on *Daphnia* heartbeat observation, refer to pp. 267–268 of Morholt and Brandwein.*

Teaching the Activity

Practice Experiment

1. Demonstrate beetle dissection, then have students perform the practice experiments according to their lab directions.

   SAFETY NOTE: Caution students not to breathe the ethyl acetate directly. Make sure the room is well ventilated and that killing jars are kept away from an open flame. Remind students to be careful when handling scissors. Instruct students as to the proper method of disposal of the dissected beetle. Refer also to general safety guidelines in the project overview.

2. Compare results as a class.

Student Designed Experiment

1. Have each student answer the questions in the student section of the lab prior to writing their hypotheses. Provide help as needed. Answers to student questions are given below.

2. Allow all students who have chosen to test their products using this lab to work on experiments individually. Provide help as needed.

Answers to Questions on Student Worksheets

1. a. and b. Most water soluble substances are also soluble in Ringer's solution, at least in low concentrations. Some water soluble substances such as vitamins have fillers

---

* (Excerpts on *Daphnia* heart rate adapted from *A Sourcebook for the Biological Sciences*, third edition, by Evelyn Morholt and Paul S. Brandwein, copyright © 1986 by Harcourt Brace Jovanovich, Inc., reprinted by permission of the publisher.)

such as starch that will settle to the bottom; these can be filtered out. Fat soluble material will be immiscible in Ringer's. These substances may be made soluble by first dissolving them in a small amount of ethanol, then dissolving this in Ringer's. Allow as much of the ethanol as possible to evaporate; avoid heating. The presence of the alcohol creates an additional variable. This can be controlled by treating a separate beetle with Ringer's containing an equal concentration of alcohol with no test substance, evaporated in the same way.

SAFETY NOTE: Refer to the cautions in the project overview.

1c. To determine the final concentration of substance the beetle heart will receive, follow these steps:

1. Determine the volume of a drop by counting the number of drops in 1 ml, then dividing 1 ml by the number of drops. Repeat this three times and take an average.

2. Determine the volume of Ringer's solution required to cover a beetle heart by pouring enough to cover the wax in the Petri dish. Measure the volume of the fluid.

3. Make up a solution of substance with known amounts of Ringer's and test substance—for example, 100 mg/100 ml = 1 mg/ml.

4. Find the amount of substance in one drop of solution using calculations from step 1 and the amount you decided on for step 3. For example:

$$\frac{X \text{ mg in 1 drop}}{0.1 \text{ ml}} = \frac{1 \text{ mg}}{1 \text{ ml}} \text{ (the concentration decided on in step 3)}$$

(0.1 ml is a possible volume from step 1)
Cross multiply and solve for x.

5. Determine the final concentration the beetle heart will receive from each drop by dividing your answer to step 4 by the answer you get in step 2 plus the volume of a drop (the answer to step 1). For example:

$$\frac{0.1 \text{ mg (answer to step 4)}}{5 \text{ ml} + 0.1 \text{ ml}} = 0.02 \text{ mg/ml}$$

(5 ml + 0.1 ml is the answer to step 2 plus answer to step 1)

The concentration may be increased by applying more drops, one at a time. Each time, be sure in the calculation for step 5, to multiply your answer to step 4 by the number of drops added and to increase the total volume in the denominator by the number of drops you added.

1d. Answers will vary.

1e. Variables may include: presence or absence of test substance, varying concentrations of test substance, or presence or absence of ethanol required to dissolve the test substance.

1f. The control experiment is the base-line heart rate measured for the beetle prior to adding the test substance. The experimental is the heart rate measured on the same beetle with the test substance.

2. A sample hypothesis might be:
   "If a beetle heart is treated with bee pollen in a concentration of 0.05 mg/ml after a base-line heart rate has been determined, then the heart rate will decrease."

3. Answers will vary somewhat, but should generally follow the practice experiment guidelines.

4. Students should set up their own data table; it may resemble the "practice data table" on p. 153.

5. Answers will vary.

Follow-up Activities

See project overview.

Evaluation

Method of Evaluation	Goal(s) Tested
1. Completion of student-designed laboratory	1,2,3,4

References

See project overview.

WHAT ARE THE EFFECTS OF SUBSTANCES ON THE HEART OF THE BEETLE, *PASSALUS CORNUTUS*?*

Practice Experiment

The beetle heart is easy to dissect and clearly visible through the dissecting microscope. It is possible to determine the effects of substances on the beetle heart by measuring whether the heart rate increases or decreases after they are added.

Dissecting Instructions

1. Remove a beetle from the container by grasping the abdomen with your fingers, staying clear of the anterior (forward) appendages, which will pinch. The beetles will make a hissing sound; this is normal.

2. Open the killing jar. Hold the beetle over it with one hand; with the other, sever the beetle at the "waist"—the area behind the prothorax—using scissors. Let the anterior (front) portion drop into the killing jar. Quickly replace the lid on the jar.

   SAFETY NOTE: Make sure the room is well-ventilated when you are using the killing jar, and do not breathe its fumes directly. Keep the jar away from an open flame.

3. Place the beetle abdomen, with legs still attached, on the dissecting pan so that the dorsal side is up.

4. Remove the elytra, or wing covers, shown in Figure 16–1, by lifting and clipping them at the base with scissors. Then remove the hind wings, taking care not to puncture the heart below.

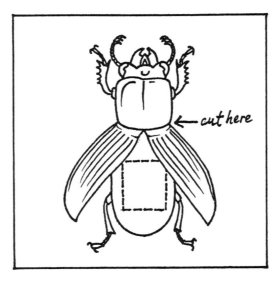

FIGURE 16–1

5. Carefully remove the heart by cutting out a "square" around the dorsal (top) abdominal wall to which the heart is attached. Lift out the square (which contains the heart) and remove and discard any attached intestine. The beetle abdomen may now be discarded.

* Beetle experiment from K. M. Wilbur and E. A. McMahan, "Low Temperature Studies on the Isolated Heart of the Beetle, *Popilius disjunctus* (Illiger)," *Annals of the Entomological Society of America*, 51, No. 1, 1958. Reprinted with permission from *Annals of the Entomological Society of America*, © 1958, Entomological Society of America.

WHAT ARE THE EFFECTS OF SUBSTANCES ON THE HEART OF THE BEETLE, *PASSALUS CORNUTUS*? (continued)

6. Place the tissue containing the heart, dorsal side up, on a wax-filled dish. Quickly flood with a premeasured volume of Ringer's solution that has been allowed to warm to room temperature.

7. Pin the tissue at the four "corners" and observe it under a dissecting microscope.

8. You should be able to see the heart beating. It is a narrow tube running down the center of the abdominal wall. When it beats, the entire tube contracts, appearing to become narrower, then wider, as shown in Figure 16–2.

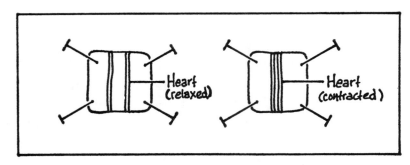

FIGURE 16–2

9. Wait several minutes until the heartbeat has become regular and stabilized. This time is necessary to allow it to recover from the trauma of dissection.

10. Establish a base heart rate by counting the number of beats per minute. Repeat this process until your number is fairly constant (within 2 or 3 beats) three times in a row. Record this number on the practice portion of your data sheet.

11. Now you are ready to begin the experiment. Apply a substance your teacher has prepared in Ringer's solution using a Pasteur pipette. Begin counting beats per minute. Take an average of three one-minute counts. Record this information on the data table.

12. To increase the concentration of the test substance, add more of it, drop by drop, to the heart. Record the information on the data table.

13. If the heartbeat becomes irregular or stops, this may be an indication that too much of the test substance has been added. You may be able to "revive" the heart by flooding it with Ringer's solution, thereby diluting the concentration of the substance.

14. The heart should continue beating for at least half an hour. When you are finished, discard the heart as directed by your teacher. Wash materials and return them to the proper area.

15. Complete the Practice Data Table.

Practice Data Table

Substance being tested:						
Base heart rate	Heart rate after adding substance			Heart rate after increasing concentration (optimal)		
	#1	#2	#3	#1	#2	#3
	average:			average:		

Did the heart stop beating? When? Were you able to revive it? How?

WHAT ARE THE EFFECTS OF SUBSTANCES ON THE HEART OF THE BEETLE, *PASSALUS CORNUTUS*?

1. You have practiced the beetle heart technique with a known substance, and you are now ready to design and carry out your own experiment. Before you write your hypothesis, answer these questions (on a separate piece of paper).
   a. Is your product soluble in Ringer's solution? You may have to try dissolving it to find out.
   b. If the answer to question a is *no*, how will you dissolve the substance in Ringer's solution? Do you need another control for this procedure? If so, what will it be?
   c. How will you determine the final concentration the beetle heart will receive? Will you use more than one concentration?
   d. What effect do you think your substance will have on the beetle heart? Will it increase or decrease its rate? On what information do you base your assumption?
   e. List the variable(s) in the experiment.
   f. Describe the control and experimental group(s).
2. Write your hypothesis.
3. Briefly list the steps of your experimental procedure.
4. Set up your own data table, and record your data on it.
5. Did the results support or refute your hypothesis? How do you explain the results, if different from what you predicted?

ACTIVITY 17

WHAT IS THE VITAMIN C CONTENT OF AN UNKNOWN SUBSTANCE?

Goals

This activity will enable students to:

1. test the claims of "health store" products;
2. be more familiar with the ingredients, modes of action, and potential benefits and dangers of some of these substances; and
3. make more informed choices about whether to use these products.

   NOTE: For synopsis, relevant topics, and age/ability levels, see the project overview on page 140.

Entry Skills and Knowledge

Before participating in this activity, students should be able to:

1. describe briefly the processes of human digestion and circulation;
2. measure liquids safely in the laboratory;
3. list some biological molecules and their functions.

Materials

indophenol (2,6-dichlorophenol indophenol)
vitamin C tablets
orange and other fruit juices
test tubes
test tube racks
Pasteur pipettes with bulbs

DIRECTIONS FOR TEACHERS

Preparation

Indophenol solution: Dissolve 1 g of indophenol in 1 l of distilled water to make up a 0.1% solution.*

Vitamin C solutions (known and unknown):

known solution: Use the values of commercial vitamin C tablets to determine the numbers of milligrams of vitamin C. The final concentration should be 1 mg/ml, for easy calculation. *Note*—Be sure your supply of "known" vitamin C is fresh, and that it has been protected from light and extremes in temperature. Otherwise, its true concentration may not reflect that on the label.

---

* Excerpt for indophenol solution adapted from *A Sourcebook for the Biological Sciences*, third edition, by Evelyn Morholt and Paul S. Brandwein, Copyright © 1986 by Harcourt Brace Jovanovich, Inc. Reprinted by permission of the publisher.

unknown solutions: You may want to let different groups determine the vitamin C content of several unknowns for comparisons. Suggested unknowns include vitamin C solution, made as the known solution was made, but in a different concentration; fresh orange juice; frozen orange juice; and tomato, apple, or grapefruit juice.

Teaching the Activity

Practice Experiment

1. Review safety notes with students (see project overview).
2. Have students perform the practice experiment according to directions on their sheets, providing help as needed.
3. Indicate to students what the unknowns were, if they haven't already guessed, and ask them to compare the vitamin C content of these. Questions you might ask include:
   - Why did fresh juice have more vitamin C than frozen (assuming it did)?
   - Which juice had the most vitamin C? How might you account for this fact?
   - To receive the greatest amount of vitamin C in your diet, which of the unknowns should you consume on a regular basis?

Student-Designed Experiment

Be sure students answer the questions in the student directions before they write their hypotheses. Provide help as needed. Answers to questions are given below. For question 4, have enough test tubes available so that students can experiment on various treatments. It is your choice whether to make up solutions for students or to have them make them up themselves.

Answers to Questions on Student Laboratory Sheets

1. Answers will vary, but may include product label information.
2. Answers will vary.
3. Answers will vary. An explanation for the difference may be the age of the product, the conditions under which it was stored, or both.
4a. An aqueous solution may be easier to expose to the treatment. Students can choose their own concentrations.
4b. Students must experiment for themselves to determine treatment conditions such as temperatures and times.
5. No, providing the same amount and concentration of indophenol are used.
6. Answers will vary.
7. A control should be run with an equal volume of water to provide a basis of comparison for the color change. Experimental group answers will vary. A sample hypothesis might be:
   > If vitamin C tablets are exposed to ambient room light for 48 hours, then the amount of vitamin C will be lower in these tablets as compared with control tablets not exposed to light.

8. Students should set up their own data tables, but they may use the table in the practice experiment as a guide.
9. Answers will vary.

Follow-up Activities

See project overview.

Evaluation

Method of Evaluation	Goal(s) Tested
1. Completion of student-designed laboratory	1,2,3

References

See project overview.

WHAT IS THE VITAMIN C CONTENT OF AN UNKNOWN SUBSTANCE?

Practice Experiment

Indophenol is a chemical that normally appears blue in a solution of water. When vitamin C is mixed with it, the solution becomes clear. By comparing the known amount of vitamin C required to cause a given amount of indophenol to change color to the amount required of an unknown solution, it is possible to obtain a rough estimate of the amount of vitamin C in the unknown substance. Perform the practice experiment as described below, then design your own.

Procedure

Record all results in Table 17–1 at the end of this worksheet.

1. Fill six test tubes each with 10 ml of 0.1% indophenol solution.
2. Label three of the tubes "known," three "unknown."
3. Determine the volume of a drop by counting the number of drops in 1 ml, then dividing 1 ml by the number of drops. Repeat the procedure three times and take an average.
4. Add the "known" solution of vitamin C to the proper tubes drop by drop, mixing well by swirling water after each drop.
5. Record the number of drops required for each tube to change the color of the solution. Average these numbers. Determine the required amount of the known vitamin C solution by the following calculations. Your teacher will provide the concentration of the "known" vitamin C solution.

   a. number of ml/1 drop × number of drops required = number of ml required

   b.
   $$\frac{x \text{ mg}}{\#\text{ml required}} = \frac{1 \text{ mg}}{1 \text{ ml}}$$
   (concentration required to bleach the indophenol)　　(known concentration)

   Cross multiply and solve for x.

For example, if it takes three drops of known vitamin C solution to bleach the indophenol, and there are 0.1 ml/drop, with a concentration of 1 mg/ml, the calculations would be:

a. 0.1 ml × 3 = 0.3 ml

b.
$$\frac{x \text{ mg}}{0.3 \text{ ml}} = \frac{1 \text{ mg}}{1 \text{ ml}}$$
(concentration required to　　(known concentration)
bleach indophenol)

In this example, x = 0.3 mg known 1 mg/ml Vitamin C substance is required to change the color of 10 ml of indophenol.

6. Now, using information from step 5, determine the concentration (per 1 ml) of vitamin C in the test substance your teacher has prepared by the following calculations. In the example

WHAT IS THE VITAMIN C CONTENT OF AN UNKNOWN SUBSTANCE? (continued)

below, it is assumed that 2 ml of unknown substance are required to bleach the indophenol, and that these 2 ml contain 0.3 mg of vitamin C, the amount calculated in Step 5. Therefore, the concentration in mg/ml of unknown test substance is:

$$\underset{\substack{\text{(amount of test sub-} \\ \text{stance required to} \\ \text{bleach indophenol)}}}{\underset{\text{2 ml}}{\underline{\underset{\text{(calculated from step 5)}}{0.3 \text{ mg}}}}} = \underset{\substack{\text{(concentration of un-} \\ \text{known substance, in} \\ \text{mg/ml)}}}{\underset{\text{1 ml}}{\underline{x \text{ mg}}}} = \underset{(2x = 0.3; x = 0.15)}{0.15 \text{ mg/ml, final concentration}}$$

7. Wash all test tubes and return materials to the proper area.
   Known substance concentration: _____

TABLE 17–1

solution	tube #	number of drops required	average	concentration
known	1			
	2			
	3			
unknown	1			
	2			
	3			

WHAT IS THE VITAMIN C CONTENT OF AN UNKNOWN SUBSTANCE?

Student-Designed Experiment

In the practice experiment, you determined the concentration of an unknown solution of vitamin C by using information about a known solution. Now you are ready to design and carry out your own experiment. Before writing your hypothesis, answer these questions:

1. What evidence indicates that your product is likely to contain vitamin C? _____

_____

_____

2. Describe briefly the purpose of your experiment. For example, do you plan to determine whether the amount of vitamin C in a commercial tablet is indeed the advertised concentra-

tion? _____

_____

_____

3. If the vitamin C concentration is found to be lower than that advertised, how might you

explain this? _____

_____

4. Do you plan to treat your product—by freezing or microwaving, for example—to assess the effect of the treatment on it? If so,
   a. How will you treat it: using dry tablets of the product, or a solution made with water? If a solution is made, what will be the concentration?

   _____

   _____

   b. Describe the exact treatment(s) the product will receive. _____

   _____

   _____

5. As you carry out this experiment, it will be necessary to use the known vitamin C concentration calculated in the practice experiment. Do you need to determine this again as you

run your experiment? _____

_____

WHAT IS THE VITAMIN C CONTENT OF AN UNKNOWN SUBSTANCE? (continued)

6. Describe your experimental variable(s). _____

_____

_____

_____

_____

_____

7. Describe your control and experimental group(s). _____

_____

_____

Write your hypothesis: _____

_____

_____

8. Set up a data table on your own paper. Be sure to indicate the treatment used, if any. Perform your experiment and record the results in your data table. Then, answer question 9.

9. Did your experiment support or refute your hypothesis? How do you explain the results,

   if different from what you predicted? _____

_____

_____

_____

_____

ACTIVITY 18

WHAT TYPES OF BIOLOGICAL MOLECULES ARE PRESENT IN SUBSTANCES?

Goals

This activity will enable students to:

1. test the claims of "health store" products;
2. be more familiar with the ingredients, modes of action, and potential benefits and dangers of some of these substances; and
3. make more informed choices about whether to use these products.

NOTE: For synopsis, relevant topics, and age/ability levels, see project overview.

Entry Skills and Knowledge

Before participating in this activity, students should be able to:

1. describe briefly the processes of human digestion and circulation;
2. measure and heat liquids safely in the laboratory; and
3. list some biological molecules and their functions.

Materials

general:

student worksheets: "What Types of Biological Molecules Are Present in Substances? Questions and Data Table" and
"What Types of Biological Molecules Are Present in Substances? Procedures"

test tube racks
test tubes (15 ml or 20 ml size)
beakers
test tube clamps
hot plates or Bunsen burners with ring stands and supports
matches or flints, if Bunsen burners are used
safety goggles
marking pencils
marking tape
Pasteur pipettes with bulbs

Chemical Test a: starches

iodine (Lugol's solution)
droppers or Pasteur pipettes with bulbs
10% soluble starch solution
porcelain spot plates

Chemical Test b: mono- and disaccharides

distilled water
Benedict's solution
10% glucose solution
10% sucrose solution

Chemical Test c: proteins

10% NaOH solution
1% $CuSO_4$ solution
meat tenderizer
distilled water

Chemical Test d: lipids

Sudan III solution
95% ethanol
filter paper
large diameter Petri dishes
forceps
distilled water
vegetable oil

DIRECTIONS FOR TEACHERS

Preparation

Here are suggestions for dissolving products, should students need assistance:

- Make up a 10% solution of each product which is soluble in water.
- For substances insoluble in water, solubilize them in ethanol first. Allow as much of the ethanol as possible to evaporate; avoid heating.

Chemical Test a: Iodine Test

1. Make up a 10% soluble starch solution, commercially available from a chemical supply house, by mixing 100 g starch per 1 l of total volume of solution.
2. Prepare Lugol's iodine solution* by:

   —mixing 30 g potassium iodide in 500 ml of distilled water;

   —adding 20 g of iodine; and

   —dispensing the solution into small bottles with droppers.

   SAFETY NOTE: Iodine stains skin and clothing. It is also corrosive, and its vapors are toxic. Avoid contact with skin, clothing, and eyes. Carry out all procedures involving iodine in a fume hood. Avoid direct heating of iodine. As with all laboratory chemicals, avoid ingesting internally.

---

* (Excerpt on Lugol's solution adapted from *A Sourcebook for the Biological Sciences*, third edition, by Evelyn Morholt and Paul S. Brandwein, copyright © 1986 by Harcourt Brace Jovanovich, Inc; reprinted by permission of the publisher.)

Chemical Test b: Benedict's Test

1. Make up Benedict's solution according to package directions.
2. Make up 10% solutions of glucose and sucrose by mixing 100 g each/1000 ml total volume of distilled water solution. Table sugar is a relatively pure and inexpensive source of sucrose.
3. Set up beakers of water on hot plates or Bunsen burners with ring stands and supports. Do not turn the heat on yet.
   SAFETY NOTE: Avoid contact of Benedict's solution with skin, clothing, and eyes. Do not inhale or ingest it.

Chemical Test c: Biuret Test

1. Prepare Biuret reagent* by making up:
   —a 10% solution of NaOH, 25 ml per student (a 10% solution means 10 g NaOH per 100 ml total volume of distilled water solution)
   —a 0.5% solution of $CuSO_4$ (0.5 g $CuSO_4$ per 100 ml distilled water solution)
2. Prepare a 10% solution of meat tenderizer in distilled water.
   SAFETY NOTE: Avoid contact of Biuret reagent with skin, clothing, and eyes. Do not inhale or ingest it. NaOH is a strong base and can cause serious damage to skin and eyes. Flood immediately with water if contact occurs.

Chemical Test d: Sudan III Test

- Make up a 2% solution of Sudan III* in absolute ethanol. Before using, dilute this with equal parts of 45% ethanol.
  SAFETY NOTE: Keep ethanol away from open flames. Avoid inhalation, ingestion, and contact with skin or eyes. Do not heat directly. Store Sudan III in metal can or glass stoppered bottle. Sudan III can stain clothing and skin, so avoid contact.

Teaching the activity

Chemical Test a: Starches

1. Have students perform tests according to the directions on their lab sheets.
   SAFETY NOTE: Remind students of general lab safety procedures *and* of precautions regarding the handling of iodine, described in the preparation section of the teacher directions.
2. For the practice test, instruct students to swirl the starch solution gently before pouring it into their own test tubes.
   Results of student tests will vary for this and all tests, as will answers on their worksheets.
   NOTE: Some over-the-counter vitamins and pain remedies, such as aspirin, contain starch as a filler. A positive iodine determination for these types of products is, therefore, not uncommon.

---

* Excerpts on procedures for Biuret, Benedict's, and Sudan III tests adapted from *A Sourcebook for the Biological Sciences*, third edition, by Evelyn Morholt and Paul S. Brandwein, copyright © 1986 by Harcourt Brace Jovanovich, Inc. Reprinted by permission of the publisher.

Chemical Test b: Mono- and Disaccharides

Have students perform tests according to the directions on their lab sheets.

SAFETY NOTE: Remind students to wear goggles at all times liquids are being heated. Make sure they do not touch the hot plates or Bunsen burner and ring stand setups directly while they are hot. Remind them to handle test tubes only with test tube clamps.

Chemical Test c: Proteins

Have students perform tests according to the directions on their lab sheets.

SAFETY NOTE: Caution students regarding safe handling of NaOH and Biuret reagent, described in the preparation section of the teacher directions.

Chemical Test d: Lipids

1. Have students perform their tests according to directions on their lab sheets.
2. Instruct students to allow each spot to dry before immersing it in the Sudan III. Students will probably be surprised to find that some vitamins (A, D, E, K) are fat soluble.

   SAFETY NOTE: Caution students regarding safe handling of ethanol and Sudan III, described in the preparation section of the teacher directions.

Evaluation

Method of Evaluation	Goal(s) Tested
1. Completion of chemical tests	1, 2, 3

References:

See project overview.

WHAT TYPES OF BIOLOGICAL MOLECULES ARE PRESENT IN SUBSTANCES?

Questions and Data Table

You will perform EACH of the chemical tests below on your product, first by practicing each test, then by designing and performing your own experiments. Before you begin, however, answer the following questions. Then, write a hypothesis indicating the biological molecules you think are present in your product.

    For each of the biological molecules below, indicate whether it is a likely ingredient in your product, and describe briefly the evidence on which you base this assumption (product labels, for example).

1. starch? _____

_____

2. mono- or disaccharides? (if so, which ones?) _____

_____

_____

3. protein? _____

_____

_____

4. fat? _____

_____

_____

Hypothesis: _____

_____

DATA TABLE 18–1

CHEMICAL TEST	TEST FOR PRESENCE OF	RESULTS
a. iodine	starch	
b. Benedict's	monosaccharides disaccharides	
c. Biuret	protein	
d. Sudan III	fats	

Name _____ Date _____

WHAT TYPES OF BIOLOGICAL MOLECULES ARE PRESENT IN SUBSTANCES?

Procedures

Chemical Test a: Is Starch Present?
Iodine (Lugol's solution) turns blue or black in the presence of starch. Practice the techniques described below with soluble starch solution, then you will be ready to design and perform your own experiment.

SAFETY NOTE: Avoid contact of the skin, eyes, and clothing with iodine; it can stain the clothing and skin. Its vapors are toxic, and it is corrosive. Avoid inhaling or ingesting it. Do not heat it directly. Procedures involving iodine should be carried out in a fume hood.

Practice Test:

1. Dispense 10 ml of 10% soluble starch solution into a test tube.
2. Add 10 drops of iodine solution to the tube and swirl.
3. You should see a purple or black color, with greatest intensity on the bottom of the tube.

Product Test:

1. Now perform the starch test on your product and record your results in Data Table 18–1.

   NOTE: Here are suggestions for applying the iodine to your product:
   a. dissolve the product in an acqueous solution; or
   b. if the product is insoluble in water, apply the iodine directly to it in a porcelain spot plate.

2. Wash all materials and return them to the proper area.

Chemical Test b: Are Mono- and Disaccharides Present?

Benedict's solution is blue. In the presence of certain mono- and disaccharides such as glucose and sucrose, the color will appear as follows:

   a. no sugar—blue (unchanged)
   b. small amount of sugar—green
   c. great amount of sugar—orange or red

Practice Test:

1. Obtain three test tubes and mark them as follows:
   #1 glucose
   #2 sucrose
   #3 control
2. Add to each tube the contents specified in Table 18–2.

WHAT TYPES OF BIOLOGICAL MOLECULES ARE PRESENT IN SUBSTANCES? (continued)

TABLE 18–2

tube #	amounts		color
	8 drops	5 ml	
1	10% glucose solution	Benedict's solution	
2	10% sucrose solution	Benedict's solution	
3	distilled water	Benedict's solution	

3. Boil all tubes in a beaker of water over a Bunsen burner or on a hot plate, as directed by your teacher, until you see a color change. This should take about five minutes.

   SAFETY NOTE: Wear goggles at all times liquid is being heated in the lab. Handle test tubes only with test tube clamps. Do not touch the beaker of water, hot plate, or Bunsen burner and ring stand set-ups directly while they are hot. Avoid ingesting or inhaling Benedict's solution. Avoid contact with skin, eyes, or clothing.

4. Record the color of the tube in Table 18–2.
5. Discard the contents of the tube as your teacher directs. Wash all materials and return them to the proper area.

Product Test

1. Now perform the Benedict's test on your product and record the results on your data sheet.
2. Wash test tubes, and return materials to the proper area.

TABLE 18–3

tube #	amounts: 5 ml	+	5 ml	+	several drops	color
1	10% meat tenderizer solution		10% NaOH		0.5% $CuSO_4$	
2	distilled water		10% NaOH		0.5% $CuSO_4$	

Chemical Test c: Are Proteins Present?

Biuret reagent turns pink, purple, or blue in the presence of proteins.

Practice Test

1. Obtain two test tubes and label them as follows:
   #1 protein
   #2 control
2. Add to each tube the amount specified in Table 18–3.

   SAFETY NOTE: Exercise great caution when handling NaOH. It is caustic, and can burn skin and eyes. Follow your teacher's directions in handling it. If contact occurs, flood immediately with water. Keep $CuSO_4$ away from skin, clothing, and eyes.

3. Mix the contents and record the final color in Table 18–3.

Product Test

1. Now perform the Biuret test on your product, and record the results on your data sheet under "Chemical Test c."
2. Wash test tubes, and return materials to the proper area.

Chemical Test d: Are Fats Present?

Sudan III is a dye that stains lipids (fats) bright red.

WHAT TYPES OF BIOLOGICAL MOLECULES ARE PRESENT IN SUBSTANCES? (continued)

Practice Test

1. Obtain two pieces of filter paper. Mark each with a pencil on the edge with the letter *C* for control or *E* for experimental. On the paper marked *E*, use a Pasteur pipette to drop a small amount of vegetable oil in the center.

2. Dip each piece of filter paper separately into a Petri dish of Sudan III solution for 1 to 2 minutes with forceps.

   SAFETY NOTE: Sudan III can stain skin and clothing, so avoid direct contact. Keep it away from open flames; avoid inhaling it directly or ingesting it.

3. Remove each piece of paper and swirl in a Petri dish of distilled water for 1 or 2 minutes.

4. Describe the difference in appearance of the vegetable oil spot and the control filter paper.

Product Test

1. Now perform the Sudan III test on your product, and record the results on your data sheet.
2. Wash test tubes, and return materials to the proper area.

section 6

Controversial Issues in Biology

An important goal of science education is to make science "relevant" for students. This means that teachers should help students:

- understand the impact of scientific issues on society, and
- learn to use science to solve problems and make decisions in their own lives.

The study of controversial topics such as genetic engineering and cloning is an excellent means of accomplishing this goal.

The strategies and activities you will find here force students to consider issues from many different points of view and in terms of their own values. In the process of making decisions, no choice will be absolutely "good" or "bad." In deciding, a person must weigh the relative value of different alternatives.

How a decision is defined as "good" or "bad" depends on the values of the person involved. These will, of course, differ. The answer is not so important as is the process of decision making.

How Should These Activities Be Used?

If you are able to teach an entire unit on bioethics, the following sequence is suggested:

1. Have students consider their own values using ideas below under "Teaching Strategies—Examining Student Values."

2. Follow the discussion of values with the activities in roughly the order in which they are presented in this section. The first three activities should promote discussion of such issues as when human life begins and what characteristics define an organism as being human. Make certain that students deal early with these concepts, since they will resurface again and again in your discussions. The next two activities deal with more global issues involving the consequences of one human being's actions on another.

3. If time permits, create additional activities of your own using the suggestions under "Teaching Strategies—Examining Specific Issues," and "Guidelines for Creating Your Own Activities."

   If you are unable to teach an entire unit on bioethics, enrich your traditional biology units with the activities presented here in ways suggested below:

   - "Should Surrogate Motherhood Be Made Illegal?" can be integrated into a unit on human reproduction.
   - "Should Parents Be Allowed to Genetically Engineer Their Unborn Children?" and "Should a Premature Infant Be Cloned?" are appropriate enrichment activities for a genetics unit.

171

- "Should Baboon to Human Transplants Be Permitted?" and "Should the Transmission of the AIDS Virus Be a Criminal Offense?" can become part of a unit on anatomy and physiology. The AIDS activity is also an appropriate supplement to a unit on microbiology.

Guidelines for Teaching

Practical Considerations

1. It is vital that you keep abreast of current developments regarding controversial topics. In the case of some issues, such as genetic engineering, new information is generated almost weekly.
2. To protect yourself from criticism of parents and school officials, follow the suggestions below before you begin discussing a controversial topic, and use the information you gather to decide a) whether to incorporate controversial topics into your class, and b) which ones you will incorporate.
   - Be familiar with the standards of your community. Be aware of the kinds of issues that are likely to be accepted or rejected as topics of discussion.
   - Check with your principal to determine school and/or system-wide policies for handling such topics.
   - Make certain that your principal is aware of what you will be teaching, and that you have his or her support.
   - Obtain permission from parents, if it is required.
   - Offer to allow those students whose parents object to the discussion to complete some other type of work, such as research in the library on a related topic, so that they can receive the same credit as the other students.

Teaching Strategies—Examining Student Values

1. Precede discussions of controversial issues with an examination of specific student values. Make sure that students understand that a value is a principle or quality considered important by an individual, such as greed, altruism, honesty, fairness, selfishness, perseverence, security, vanity, wisdom, creativity, power.

   Questions to stimulate discussion include these:
   - What are your values? Where did they come from?
   - Can you truly claim that you value something if you fail to act on it?
   - To what extent do your values influence your decisions?
   - Under what circumstances should you try to change your values?
   - Can you define a value as "right" or "wrong" in all situations?
   - Can you define "right" or "wrong"? Do these definitions vary from culture to culture?
   - Are there objective criteria for judging "right" and "wrong"?
   - Can you make decisions according to what "feels right"? What are these feelings based on?

2. Other tools for looking at the development of values include Kohlberg's levels of moral reasoning (Kohlberg, 1974), and the value clarification model (Simon, Howe, and Kirschenbaum, 1972). See also recent criticisms of Kohlberg's work (Liebert, 1984), and current research on adolescent reasoning in science-technology-curricula (Fleming, 1986).

3 Use the activity presented here entitled "Whom Would You Choose to Continue the Human Race?" to facilitate a discussion of values.

Teaching Strategies—Examining Specific Issues

1 Some of these techniques for addressing specific issues are described in detail in the follow-up sections of the "controversial issues" activities in this book:
panel discussions
speakers
films and videotapes
field trips to:
—a vivarium
—a hospital
—a genetic counseling center
—a center for the treatment of genetic diseases
—a funeral home

2. Case studies and simulations are used in some of the activities presented in this section. Some of these issues may be resolved very soon; the methods can be applied to current issues, nonetheless, following the guidelines below.

Guidelines for Creating Your Own Activities:

- Read current news articles related to controversial issues. As you do, make a list of unresolved ethical and legal questions.

- Choose one of these questions. Write a hypothetical case study (see "Should a Premature Infant Be Cloned?" for an example of a case study), or apply one of the simulation methods below.

- Make three lists students can role play and/or debate the issue based on information in these lists:
—arguments in favor of the position
—arguments opposing the position
—persons likely to be affected

3. Under each simulation below are listed examples in this book and other issues around which this type of simulation might be written:

- *SIMULATION*: Write a law making _____ legal or illegal. Defend it, addressing arguments in favor of and opposed to it.
  EXAMPLES: "Should Surrogate Motherhood Be Made Illegal?" and "Should the Transmission of the AIDS Virus Be a Criminal Offense?"
  OTHER ISSUES: cloning, euthanasia, fetal surgery, human experimentation

- *SIMULATION*: a news conference
  EXAMPLE: "Should Parents Be Allowed to Genetically Engineer Their Unborn Children?"
  OTHER ISSUES: cryonics, abortion, cloning

- *SIMULATION*: A medical ethics committee of a hospital decides whether or not a procedure should be permitted.
  EXAMPLE: "Should Baboon to Human Transplants Be Permitted?"
  OTHER ISSUES: genetic engineering, fetal experimentation, fetal surgery, cryonics, euthanasia

- *SIMULATION*: a courtroom jury trial
  EXAMPLE: "Should Parents Be Allowed to Genetically Engineer Their Unborn Children?"
  OTHER ISSUES: surrogate motherhood, transplants, bionics

- *SIMULATION*: a city council meeting
  EXAMPLE: "Is the Water Around You Safe to Drink?"
  OTHER ISSUES: genetic engineering, AIDS, euthanasia, animal rights

Guidelines for Facilitating Discussion of Controversial Issues:

- Use a "round table" for discussion, or an equivalent, such as arranging desks in a circle. Students should be able to see each other as well as the teacher.

- Be a discussion leader, but not a "dictator."

- Be totally familiar with the material in advance.

- Encourage participation by quiet students, but do not force them to speak.

- Allow a free exchange of ideas.

- Do not allow one or two students to monopolize a discussion.

- Ask unbiased questions.

- Remain neutral by:
  encouraging expression of all points of view, and
  not praising "right" answers or punishing "wrong" ones.

References

1. Fleming, R. "Adolescent Reasoning in Socio-Scientific Issues, Part I: Social Cognition," *Journal of Research in Science Teaching*, *23*, no. 8 (1986), 677–687.

2. Fleming, R. "Adolescent Reasoning in Socio-Scientific Issues, Part II: Nonsocial Cognition," *Journal of Research in Science Teaching*, *23*, no. 8 (1986), 689–698.

3. Kohlberg, L., *Developmental Psychology Today*, Delmar: CRM Books, 1974.

4. Liebert, R. M., "What Develops in Moral Development?", in *Morality, Moral Behavior, and Moral Development*, ed. William M. Kurtiner and Jacob L. Gewirtz. New York: John Wiley & Sons, Inc., 1984.

5. Simon, S. B., Howe, L. W., and Kirschenbaum, H., *Values Clarification*. New York: Hart Publishing Company, 1972.

ACTIVITY 19

SHOULD SURROGATE MOTHERHOOD BE MADE ILLEGAL?

Goals

After completing this activity, students will:

1. be more familiar with arguments in favor of and against surrogate motherhood; and
2. better understand potential legal issues associated with surrogate motherhood.

Synopsis

Students will write a law making surrogate motherhood either legal or illegal and defend it to the class.

Relevant Topics

surrogate motherhood
human reproduction
artificial insemination

Age/Ability Levels

grades 10–12, average to gifted

Entry Skills and Knowledge

Before participating in this activity, students should:

1. be able to explain basic principles of human reproduction; and
2. be able to define surrogate motherhood.

Materials

student worksheet: "Should Surrogate Motherhood Be Made Illegal?"

DIRECTIONS FOR TEACHERS

Preparation

Photocopy student worksheets.

Teaching the Activity

1. Distribute student worksheets. Explain the assignment briefly, and be sure students understand what surrogate motherhood is.
2. Ask students to cite:
   - reasons people might choose to be surrogate mothers;
   - reasons a couple might choose to hire a surrogate mother; and
   - arguments for and against surrogate motherhood.

   Accept all ideas, not singling out any one as either bad or good.
3. Have students work in pairs.

4. Provide time in class for completion of worksheets. Answer questions of a factual nature, but have students come up with their own ideas regarding legal and ethical issues.

5. When all students have finished, reassemble as a class. Call for a show of hands as to how many wrote "legal" and "illegal" laws.

6. Ask students to begin explaining their laws. As they do, record the following information on the board (you can use the table below):

   • a tally of the number of "legal" and "illegal" laws;
   • circumstances cited in the laws under which surrogate motherhood should be legal or illegal;
   • arguments for legality of surrogate motherhood; and
   • arguments in favor of surrogate motherhood being illegal.

	legal	illegal
circumstances		
arguments for		

Examples of answers and arguments students might cite include the following:
Circumstances under which it might be *legal* include:

• all;
• if the adoptive parents are unable to conceive; or
• if the adoptive parents are under age 40.

Circumstances under which it might be *illegal* include:

• all;
• if the health of the surrogate mother would be impaired by having a child;
• if the surrogate mother is having the baby solely for the money; or
• if the surrogate mother is single.

Arguments for legality include:

• Some couples are unable to have children. In contrast to traditional adoption, surrogate motherhood can assure that the adoptive father is also the biological father. *Note*: A surrogate mother is often inseminated with the sperm cells of the father of the adoptive couple; in the future, however, sperm cells from other sources, such as donors to a sperm bank, may also be used.

• Fewer infants are available for adoption now than in previous years, limiting the choices for couples who are unable to conceive.

• The surrogate mother may experience a feeling of joy because she is helping a childless couple enrich their lives.

• It is a way to enable a woman who enjoys being pregnant to make a living.

Arguments against legality include:

- Problems can arise if the surrogate breaks the contract. For example:
  - —The legal ramifications would be unclear and have not been worked out yet.
  - —Which parent(s) would be granted custody of the child would have to be decided by courts.
  - —There would be a great deal of anguish for all parties involved, including the child.
- The surrogate mother might change her mind after the adoption, and attempt to regain custody or see the child against the wishes of the adoptive parents.
- A situation might be created in which only rich, childless couples could adopt children, and only poor women would become surrogates.
- Some may feel that artificial insemination of a surrogate with the sperm cells of a man she is not married to, and/or who is married to someone else is immoral.

7. Be sure all groups have been able to speak before a discussion begins. At this point, discussion will probably be spontaneous. If students have failed to list any of the arguments given above, bring these up yourself. Other questions you can cite, if students fail to, include these:

- Should the child be told about the circumstances of its birth? If so, at what age? If it is told, how might it feel about the situation?
- Might an emotional bond exist between the biological father and the surrogate mother if his sperm cells were used to inseminate her? If so, how should this be dealt with?
- Should the adoptive parents be present at the birth of the child? Explain your answer.
- Should the child be put into the arms of the biological mother immediately after birth? The adoptive mother? Explain.
- What are arguments for and against the adoptive parents and the surrogate mother knowing each other prior to the birth?
- If the surrogate mother is married, how might her husband view her being hired to have a child for another couple? How would you view this if you were the husband?
- What would happen if the child were born with a genetic disease? What are ways to assure that this does not happen? (Should the surrogate be required to have amniocentesis and an abortion if a genetic disease is found, for example?)
- What are some similarities and differences between this situation and that of traditional adoption? Explain.
- Should the surrogate be allowed to visit the child as it grows up?
- If the surrogate neglected to receive proper prenatal care, what would be the obligations of the parents to her and the child? How could this situation be avoided?
- What would happen to a severely handicapped or mentally retarded child should any of the parents—surrogate or adoptive—decide not to accept it?

Answers to Questions on Student Worksheet

Writing Your Law

1. Answers will vary; refer to "Teaching the Activity" for possible answers.
2. Answers will vary, but should follow closely what the student has written for question 1.

3. Answers will vary; refer to "Teaching the Activity" for possible answers.
4. Answers will vary; refer to "Teaching the Activity" for possible answers.

Follow-up Activities

- Have students write an application that a woman who wishes to become a surrogate mother might complete. Include a list of criteria a woman must meet to be considered a potential surrogate. Consider the importance of:

  —the mental, emotional, and physical health of the woman;

  —her reasons for becoming a surrogate; and

  —the likelihood of the person breaking the contract.

- Have students write an application to be completed by parents wishing to hire a surrogate mother. Include a list of criteria the individuals must meet to be considered parents of a surrogate child. Consider such factors as:

  —the mental, emotional, and physical health of both persons;

  —stability of the marriage;

  —their reasons for wanting a surrogate child;

  —income and socioeconomic status of the man and woman;

  —the likelihood of the individuals breaking the contract.

- As a class, discuss court cases (such as "Baby M") involving surrogate motherhood. Keep up with legal precedents made on the state and national levels.

- Have students write a contract between a set of adoptive parents and the surrogate mother. Consider such issues and responsibilities of the surrogate, including the following:

  While pregnant, should she be required to:

  —have monthly prenatal check-ups?

  —maintain a consistent diet designed by a nutrition expert?

  —avoid harmful drugs?

  —have amniocentesis?

  —have an abortion if the fetus is found to have a genetic disease?

  Consider also possible responsibilities of the adoptive parents:

  —How much money should they give the surrogate mother? What should they be required to pay her if she breaks any part of the contract?

  —Under what circumstances are they allowed not to accept the child? (for example, if the surrogate has broken her part of the contract, if the child has a deformity that was not detected prenatally?)

  Students should consider also the relationship between the surrogate, the child, and the adoptive parents after the birth of the child, and the consequences of any party breaking the contract.

- Have students cite arguments for and against each couple below being chosen as adoptive parents of a surrogate child. All profess to want children very much.

—Both the husband and wife are in their early thirties and have been married seven years. The couple can conceive, but the woman chooses not to carry a child herself because of her demanding career as a research chemist. Her husband is a professor of art at a major university.

—The husband is 35; the wife 33. Both are well educated. The woman is unable to conceive; both parents wish to spend a great deal of time at home with the child. The husband is president of a large corporation; his wife is a trained paralegal. He was an abused child, and vows that he will never treat his children the way he was treated.

—The husband is 45; the wife 41. She is able to conceive, but her doctor advises her that because of her body structure, carrying a child to full term might endanger her health. The wife is a high school chemistry teacher; the husband is a building contractor.

- Have students play roles of court jurors, deciding which of the three couples above should be allowed to hire a surrogate mother. Assume that only one surrogate is available.

- Have students cite arguments for and against each woman below being chosen as a surrogate mother:

—This woman is 26, and happily married with a child of her own. Her husband supports her decision; both want to help another couple unable to conceive. Her husband is an electrical engineer; she is a licensed paramedic.

—This woman is 23, single, and is in excellent health. A registered nurse, she has decided to take a year off before attending medical school. She needs to support herself somehow, and decides that she would rather make money as a surrogate than as a nurse.

—This woman is 35, with two healthy children of her own. She is happily married; she and her husband are both lawyers. She had a child when she was 16, before she was married, and gave it up for adoption. She feels guilty about having to give up that child and sees being a surrogate as a way to, in her words, "make up for the mistake" she made when she was young.

Evaluation

Method of Evaluation	Goal(s) Tested
1. Participation in class discussions	1,2
2. Completion of worksheets	1,2
3. Completion of activities listed in follow-up section	1,2

References

1. Kantrowitz, B., McKillop, P., Joseph, N., Gordon, J., and Turque, B., "Who Keeps Baby M?", *Newsweek,* January 19, 1987, pp 44–51.

Name _____   Date _____

SHOULD SURROGATE MOTHERHOOD BE MADE ILLEGAL?

At present, surrogate motherhood is legal in some states, but not in others. Many couples are hiring surrogate mothers as an alternative to adoption. They pay a surrogate up to $40,000 plus medical expenses to carry a child conceived with the adoptive father's sperm cells. Recently, legal questions regarding the issue have surfaced as some surrogate mothers have chosen to break the contracts.

A Surrogate Motherhood Law: Background

1. Write a law making surrogate motherhood either legal or illegal in your state. Under "Writing Your Law" below, include the following:

   *** the circumstances under which it should be made legal or illegal (For example, should the adoptive mother's ability to have children be a factor? Should the age of the adopted parents or motives of the surrogate mother be relevant?)

   *** arguments supporting your law, and

   *** arguments others might cite opposing your law.

2. Be prepared to present and defend your law to the class.

Writing Your Law

1. Does your law make surrogate motherhood legal or illegal? Under what circumstances?

   _____

   _____

   _____

2. Write your law: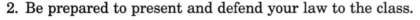

   _____

   _____

   _____

   _____

SHOULD SURROGATE MOTHERHOOD BE MADE ILLEGAL? (continued)

3. List arguments supporting your law:

a. _____

b. _____

c. _____

d. _____

e. _____

4. List arguments others might cite opposing your law:

a. _____

b. _____

c. _____

d. _____

e. _____

ACTIVITY 20

SHOULD PARENTS BE ALLOWED TO GENETICALLY ENGINEER THEIR UNBORN CHILDREN?

Goals

After completing this activity, students will:

1. be more familiar with the potential benefits and dangers of genetic engineering; and
2. be aware of arguments for and against genetically engineering a new human being.

Synopsis

Students will simulate a court case deciding whether "short" parents are allowed to genetically engineer a "tall" child. A mock news conference will also be held.

Relevant Topics

genetic diseases
recombinant DNA
protein synthesis
interactions between genes and environment

Age/Ability Levels

grades 10–12, average to gifted

Entry Skills and Knowledge

Before participating in this activity, students should be able to:

1. explain what recombinant DNA is;
2. describe how certain genetic diseases are transmitted; and
3. list the symptoms of some genetic diseases.

Materials

student worksheets: "Should Parents Be Allowed to Genetically Engineer Children? Background Information" and "Should Parents Be Allowed to Genetically Engineer Children? Worksheet"

DIRECTIONS FOR TEACHERS

NOTE: Roles may be added or omitted as the size of your class dictates. For example, you may add members of "Right to Natural Conception." The news conferences can be omitted if there are too few students. In a real court case, some of the jurors listed would probably not be chosen. Their points of view, however, can help create interesting discussions. Tell students to pretend that these jurors were chosen, for the sake of the activity.

Preparation

1. If possible, have students read a current article on genetic engineering.
2. Photocopy student worksheets.

Teaching the Activity

1. Distribute worksheets to students. Allow them time to read the case.

2. List all roles on the board and allow students to choose. Cross them off as they are taken.

3. Explain the procedure to students:

   a. News reporters will hold a pretrial press conference interviewing any participants in the simulation except the jurors.

   b. Because of time constraints, only the closing arguments of lawyers will be presented, without interruption, to jurors. Jurors should take notes during this phase.

   c. Jurors will discuss the arguments among themselves, with all other class members present, then vote on the case.

   d. One of the jurors will read the verdict, with an explanation. If a unanimous or majority vote did not occur, the juror will indicate this and briefly summarize arguments on both sides.

   e. The news reporters will conduct a "post-trial" interview with all trial participants, including jurors.

4. Give students time in class to prepare their roles. Allow:

   • each lawyer to work with his or her clients in preparing arguments;

   • the jurors to discuss reasons each member might be in favor of or opposed to the case, and how these viewpoints might affect the outcome of the case;

   • the reporters to work together in writing interview questions.

5. Monitor the news conferences and trial, then hold a class discussion. Cite the following arguments for and against the case, if they have not already been mentioned.

For

• Most parents wish to give their children the best possible advantages in life.

• Any work involving recombinant DNA will contribute to our knowledge of how genes are regulated. Understanding gene regulation in higher organisms may enable humans to:

—regenerate limbs

—slow down or accelerate the aging process

—cure cancer

• Genetically engineering an unborn child may make it possible to cure genetic diseases such as sickle-cell anemia, hemophilia, and Tay-Sachs disease.

Against

• Some scientists argue that billions of years of "evolutionary wisdom" have given us existing gene combinations. In changing these, we are tampering with this "wisdom," without understanding the evolutionary consequences of these new combinations. (A counter argument to this is that evolutionary "wisdom" has given us such diseases as AIDS and cancer. Keeping people alive using techniques of modern medicine runs counter to evolutionary "wisdom;" does this mean, then, that people should no longer be treated in hospitals?)

• Scientists do not yet fully understand how genes in eukaryotes are regulated. The position of a gene in relation to other genes may very well affect its expression. If the positions of genes are changed, we may deregulate potentially harmful genes which formerly were silent. In the same way, genes which formerly were expressed may be rendered inactive.

- Mixing genes from existing organisms may create new and unpredictable combinations of proteins, some of which may be harmful or lethal.

- There is a remote danger that the procedure will not work, producing a genetic "monster." If this happens, who is responsible, and what will be done with it?

- It is unethical to change the characteristics of a person as yet unborn; "nature" should be allowed to "take its course."

6. Other questions to stimulate discussion include:

- How might the decision of the jury have changed, if at all, in each of the situations below?

  Two parents consider themselves "ugly," and want to have a physically "beautiful" child.

  Both parents carry the gene for a recessive fatal genetic disease that at present has no cure. They want the abnormal genes replaced with normal ones.

  Two parents are both "geniuses," with IQs above 160. They have always been stigmatized and called "eggheads." They want their children to be of "average" intelligence, with IQs around 100.

  Scientists want to create a new human carrying the gene for chlorophyll in the skin, so it will be able to manufacture its own food in sunlight.

  Scientists want to create a human being with very keen aural ability, using the "hearing" genes from a whale.

- Will the proposed engineering of human traits always be voluntary? If not, who will control the process, and for what purposes?

- Will the person being "created" resent having had his or her fate determined by others? Is not changing the traits of a person a statement that the original characteristics were "not good enough?"

- To what extent are traits such as height, mental ability, and aggression determined by genes? To what extent by environment? Will genetic engineering alone assure that a particular characteristic will be expressed as planned in the adult? Why or why not?

- To what extent are our physical and mental characteristics already determined for us? In what ways? Are "normal" nonengineered people already robbed of their freedom to become whatever they wish, due to such influences as family and societal values, in the same way "engineered" people might be said to be robbed of their freedom?

- Is it ethical to change the traits of persons as yet unborn? Explain.

- What would be the long-term effects on society of:

  curing all genetic diseases, including cancer?

  —eliminating mental retardation?

  —controlling the aging process?

  —eliminating all birth defects?

- If genetic engineering of unborn children were a routine procedure, could a 16-year-old child with a physical deformity sue his parents for not having taken advantage of genetic engineering procedures that have proven to be successful in correcting the genes that caused his deformity?

Answers to Questions on Student Worksheets

1. and 2. Answers will vary.
3. and 4. Answers will vary; for possible arguments, refer to "Teaching the Activity."

> 5. Answers will vary. The parents might be asked to explain their reasons for wanting to genetically engineer their child; the hospital employees and "Right to Natural Conception" group might be asked to explain their arguments in support of or in opposition to the case.

Follow-up Activities

- Have students choose one "desirable" human trait to proliferate, or one "undesirable" one to eliminate through genetic engineering.
- Describe:
  —the purpose of the procedure;
  —ethical questions it raises; and
  —scientific consequences to the rest of society.

- Have them present their ideas to the rest of the class.

Some traits which might be engineered include:

> changing mental ability (Would "superior" people advance society? Would "inferior" ones not mind performing menial labor?)
>
> changing physical characteristics
>
> enhancing creativity
>
> manipulating emotional characteristics
>
> eliminating a genetic disease
>
> slowing or accelerating the aging process
>
> **also:**
>
> a new life form could be created consisting of a human in combination with "desirable" genes from another organism

- Have students keep up with new scientific advances and legal precedents in the area of genetic engineering.
- Have the class listen to a panel discussion on whether a case such as this one is plausible in the near future. Be sure to prepare questions in advance. Members of the panel might include:

  genetic researchers

  doctors

  hospital administrators

  nurses

  genetic counselors

  lawyers

  persons with genetic diseases, or their family members

- Have one of the persons listed above speak to the class about how the genetic engineering of fetuses could change the person's eventual profession or lifestyle.

● Repeat this activity with one of the scenarios of the first question in #6 under "Teaching the Activity."

Evaluation

Method of Evaluation	Goal(s) Tested
1. Participation in the simulation exercise	1,2
2. Completion of student worksheet	1,2
3. Participation in class discussion	1,2
4. Question 1	2
5. Question 2	1
6. Question 3	1,2

Question 1

List three arguments in favor of genetically engineering a new human being, and three arguments against the process.

Question 2

List two possible benefits of genetically engineering human beings, and two possible dangers.

Question 3

Describe a situation in which parents might want to genetically engineer their unborn child. Cite arguments for and against their decision.

SHOULD PARENTS BE ALLOWED TO GENETICALLY ENGINEER THEIR UNBORN CHILDREN?

Background Information

You are going to simulate a court case in which you decide whether two parents are allowed to use genetic engineering to choose the characteristics of their unborn child. There will also be news conferences about the case. The case and roles from which you are to choose are described below. Use the accompanying worksheet to plan your strategy. Your teacher will give you further directions.

Case

A husband and his wife are both in their early thirties. The woman is 4 feet 8 inches tall and the man, 5 feet 1 inch tall. All of their lives, the two of them have been called "shorty" by others. Both were excluded from sports in school, especially basketball, because of their height. The man has recently learned that he was passed over for a job promotion because his boss thought that "only someone of larger stature could command the respect that a job with that level of responsibility demands."

Now they would like to have a child. Although they realize that height should not be the problem that it is, they do not want a child to experience the same difficulties in life that they have had to face. Instead of passing on their "short" genes, they want to give the child a chance to be taller than they are. They have read about a new process in a local hospital in which genetic engineering can create a child of desired characteristics. All of the genes of the parents are used except those that are "undesirable." In this case, the "short" genes will be replaced by "tall" genes from a bank of donated reproductive cells maintained in a hospital laboratory. Persons who donated these cells signed a consent form giving the hospital permission to use their genes in this way.

A newly formed organization, "Right to Natural Conception" group, is suing the parents and the hospital, claiming that such "tampering with nature" is immoral and should be illegal.

Roles

NOTE: For the role you choose, you may fill in details not listed, such as occupation, socioeconomic status, and age.

Presenters:

- "Right to Natural Conception" side:

  —a lawyer

  —four members of the "Right to Natural Conception" group who assist the lawyer in preparing the case
- Parents' side:

  —a lawyer for the parents and the hospital

  —one set of parents

  —two doctors

  —two hospital administrators

SHOULD PARENTS BE ALLOWED TO GENETICALLY ENGINEER THEIR UNBORN CHILDREN? (continued)

Jurors:

- a basketball superstar, 6 feet 11 inches tall
- a man who is a chemical engineer, 5 feet 5 inches tall
- a woman, 28, who may have inherited Huntington's Disease
- a man with sickle cell trait
- the mother of a child with Down's syndrome
- a man with diabetes
- a woman with heart disease
- a man born with severe facial deformities
- a woman carrying the gene for a recessive fatal genetic disease
- a nurse who works in the maternity ward of a hospital
- an obstetrician
- a clergy member, denomination to be chosen by you

News Reporters:

- reporter #1
- reporter #2

SHOULD PARENTS BE ALLOWED TO GENETICALLY ENGINEER THEIR UNBORN CHILDREN? (continued)

Worksheet

1. What role have you chosen? _____

   (*If you are a news reporter, skip to #5.*)

2. Is this person likely to be for or against the decision of the parents? Explain. _____
   _____

3. List arguments you will make in favor of your position:

   a. _____
   _____

   b. _____
   _____

   c. _____
   _____

4. List arguments others might cite in opposition to your position. Beside each, indicate how you will address it.

   a. _____   _____
   _____   _____

   b. _____   _____
   _____   _____

   c. _____   _____
   _____   _____

For news reporters only:

5. List the questions you will ask:
   a. the parents: _____
   _____

   b. the employees of the hospital (doctors and administrators): _____
   _____

   c. members of the "Right to Natural Conception" group: _____
   _____

ACTIVITY 21

SHOULD A PREMATURE INFANT BE CLONED?

Goals

After completing this activity, students will:

1. be aware of arguments for and against the cloning of human beings; and
2. be more familiar with the possible consequences of cloning.

Synopsis

Students will argue both sides of a hypothetical case of parents wanting to clone a premature infant.

Relevant Topics

human reproduction
anatomy and physiology
embryology
genetics

Age/Ability Levels

grades 10–12, average to gifted

Entry Skills and Knowledge

Before participating in this activity, students should be able to explain:

1. basic principles of human reproduction; and
2. how genetic abnormalities occur.

Materials:

student worksheet: "Should a Premature Infant Be Cloned?"

DIRECTIONS FOR TEACHERS

Preparation

1. If possible, have students read an article explaining the procedure of cloning.
2. Photocopy student worksheets.

Teaching the Activity

1. Distribute worksheets and give students time to read them.
2. Be sure all students understand what *cloning* means.
3. Give students time in class to complete their worksheets. It is your choice whether to have them work in pairs or individually.
4. When all worksheets are completed, ask for arguments supporting each statement. Record them on the board, marking checks beside each to indicate the number of times it is cited.

5. Bring up these arguments, if students fail to:

For

- Cloning is probably the only way these parents can have a child that is biologically their own.
- There are fewer infants available for adoption than there were years ago, so it may be difficult for these parents to have a child at all without cloning.
- This procedure would allow the premature infant a chance to live.
- A successful cloning procedure could contribute a great deal to our knowledge of embryological development.

Against

- It is wrong to clone a human being without its consent; the infant in this case is unable to speak for itself. Cloning in this case is the equivalent of experimental research.
- Such a procedure is unnatural, and therefore unethical.
- If the procedure is unsuccessful, would it not cause the parents even more grief?
- Would the death of the clone be considered murder? Who would be responsible?

6. Ask students which viewpoint they support and why.

7. Ask these questions to stimulate discussion:
   - Do you think cloning should or should not be allowed in each situation below? Explain your reasons.
     - Two wealthy parents want to clone their newborn healthy infant. There is a history of heart disease in their family; they fear that the infant will need a transplant later in life and want to be sure a compatible heart is available. They reason that if they clone a "replica" now and "grow" it under sedation for "spare part use" later on, the heart should be about the right size by the time it is needed by the "original" infant.
     - A brilliant 80-year-old scientist is dying. She wants to be sure that her "smart" genes remain on earth so that she can continue to benefit the human race. Her solution is to produce a clone of herself.
     - A woman who is a twin has had a child. She wants the child to have the benefit of growing up with a constant "best friend," as she did. She sees cloning as a way of providing him with a "twin" companion.
     - The president of a small country wishes to clone a group of people with characteristics he thinks would make them good soldiers—large body stature, obedience, loyalty, great physical strength, keen eyesight, and excellent hearing. He plans to use his "army" to defend his country against outside invaders.
   - Would the plans of the people in the scenarios above work? Why or why not? Before answering, consider the following:
     - When a clone is made from a human being, should the "duplicate" have the same legal rights as the "original?" Why or why not?
     - How do you think a clone would feel about how it was brought into existence? How would you feel if you knew you had been cloned?
     - Does having the same genes insure that two people will appear and behave exactly alike? To what extent is environment a factor in determining these characteristics?

—How are identical twins like clones? Unlike them? Do all "identical" twins look and behave exactly alike? If not, to what do you attribute the differences?

—Would you want to have yourself cloned? If not, why not? If so, for what reason?

—If you were a clone, how would you feel toward your "original"?

—If you were an "original," how would you feel toward your clone?

Answers to Questions on Student Worksheet

I. 1. and 2. Answers will vary. See #5 under "Teaching the Activity " for possible arguments.
II. Answers will vary.

Follow-up Activities

● Show the movie *The Boys from Brazil,* about the hypothetical cloning of Hitler by Dr. Joseph Mengele. Is such a scenario possible? If Hitler had been cloned, what might be the consequences? Follow this up with reading an article about Dr. Mengele.

● Have studentrs write a law indicating circumstances in which cloning should be made legal or illegal.

● Have students write a law indicating circumstances in which cloning should be made legal or illegal.

—Who is responsible in the event the clone is deformed, assuming the "original" was not deformed? Can the parents sue the hospital?

—What will happen to a deformed clone? Will it be raised by the parents? Raised in a foster home? Put up for adoption?

—Who will pay for the procedure? How much should be paid?

—If the clone dies before it has reached its full term, is the hospital obligated to "reclone" it, without charge to the parents?

—Will any type of experimentation be permitted on the clone, either during the cloning process, or later in life?

—Should the clone be told how it came into existence? If so, at what age, and by whom?

—Should the death of a clone be considered murder? If so, who is responsible?

● Plants can be cloned easily using kits available from biological supply houses. Clone plants to demonstrate that the process is really possible.

Evaluation

Method of Evaluation	Goal(s) Tested
1. Completion of student worksheets	1,2
2. Participation in class discussions	1,2
3. Question 1	1
4. Question 2	2

Question 1

List three arguments in favor of and three opposing the cloning human beings.

Question 2

List five possible consequences of cloning human beings.

References

1. Cooke, R., "The Search for Life After Death," *The Boston Globe*, October 17, 1976, pp. A–3 (reprinted in *Social Issues Resources Series*, Volume 1, Article 18).

2. Maranto, G., "Clone on the Range," *Discover*, August 1984, pp. 34–38.

SHOULD A PREMATURE INFANT BE CLONED?

Cloning is a process in which the chromosomes of a somatic cell are inserted into an egg cell which has had its nucleus removed, as shown below. Theoretically, an entire adult organism can be produced in this way from one cell. Although frogs and some plants have been cloned, the process has not been successful with human cells.

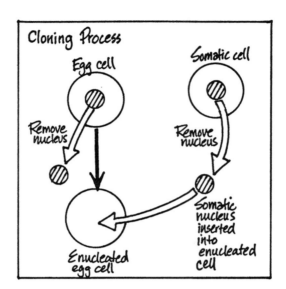

Read the hypothetical case below, then list arguments for and against cloning in this situation.

Case:

A woman, age 39, and her husband, age 43, have tried to have children since their marriage nine years earlier. The woman has had three miscarriages, and her doctors have warned her that because of her advanced age, any children born to her from now on might have birth defects.

Nonetheless, she recently delivered an infant boy at six months gestation. Although the child appeared physically normal, it was quite weak and had serious respiratory problems attributable to its early birth. It died soon after it was born, but, at the request of the parents, the doctors are keeping some of its skin cells alive in tissue culture dishes to be used for cloning.

The couple has heard about an experimental procedure at a major university in which human beings can be cloned. They desperately want a child of their own and fear that this may be their last chance. Should the parents be allowed to clone their premature infant?

I. List arguments supporting each of the statements below:

　1. Yes, the parents should be allowed to clone their premature infant, because:

　　a. ——————————————————————————————————————

　　　 ——————————————————————————————————————

　　b. ——————————————————————————————————————

　　　 ——————————————————————————————————————

　　c. ——————————————————————————————————————

　　　 ——————————————————————————————————————

SHOULD A PREMATURE INFANT BE CLONED (continued)

2. No, the parents should not be allowed to clone their premature infant, because:

   a. _____

   _____

   b. _____

   _____

   c. _____

   _____

II. Which of the above statements do you agree with? Why?

_____

_____

_____

_____

ACTIVITY 22

SHOULD BABOON TO HUMAN TRANSPLANTS BE PERMITTED?

Goals

After completing this activity, students will:

1. be more familiar with arguments for and against the use of animals as organ donors for humans;
2. understand some of the medical difficulties involved in transplantation; and
3. be aware of some of the potential psychological and emotional problems associated with transplantation.

Synopsis

Some students will play the roles of members of a hospital bioethics review board. They will listen to a case presented by other students and decide whether or not a hypothetical animal to human heart transplant is to be permitted.

Relevant Topics

transplantation
anatomy and physiology
tissue rejection
immunology

Age/Ability Levels

grades 10–12, average to gifted

Entry Skills and Knowledge

Before participating in this activity, students should be able to:

1. explain the basic anatomy and function of the human heart;
2. define the term "transplant"; and
3. be able to cite reasons for transplanting organs in human beings.

Materials

student worksheet: "Should Baboon to Human Transplants Be Permitted? Background Information" and "Should Baboon to Human Transplants Be Permitted? Preparation Worksheet"
a model of a human torso showing a human heart (optional)

DIRECTIONS FOR TEACHERS

NOTE: You may add or omit members of the review board, as the size of your class dictates. To add members, have more than one student play a given role. Be sure to have representatives from as many different walks of life as possible.

Preparation
1. photocopy student worksheets
2. obtain human torso model, if you plan to use it

Teaching the Activity
1. Distribute the student worksheets and briefly explain the simulation. If possible, have students read a news article about baboon to human transplants.
2. Ask students why some people might be opposed to a baboon to human transplantation and why others think it is necessary. Accept all responses.
3. On the board, write suggested roles given in the student directions. Cross each one off as it is chosen by a student.
4. Have students complete the preparation worksheet. Provide help as needed. You will need to let the "parents" work together. You may want to allow other students with the same roles to work together, as well.
5. Be sure that:
   - the parents decide their socioeconomic backgrounds, marital history (are they married, divorced, unmarried?), occupations, and beliefs;
   - the clergy members choose different denominations; and
   - the community members choose their occupations, socioeconomic backgrounds, and beliefs.
6. Reassemble students as a group. Explain this procedure to all students:
   a. Presenters will speak first without interruption. As board members think of questions, they should write them down.
   b. When presenters are finished, board members are allowed to ask their questions.
   c. Board members will confer among themselves as they decide the case. The presenters, however, will be present. Each board member should have a chance to speak.
   d. Board members will vote whether to permit the transplantation. Since students are playing roles and their votes do not necessarily reflect personal opinions, secret ballots should not be needed.
7. Allow at least one class period for the simulation.
8. When the decision is made, cite the following arguments for and against baboon to human transplantation, if they have not already been discussed. Ask for student reactions to them.

   For
   - Baboons are not endangered and can be bred solely for medical purposes. Therefore, unlike human hearts, a ready supply of baboon hearts is possible.
   - Human hearts for transplantation purposes are difficult to find, particularly for infants.
   - Transplantation research could be done more easily with baboon than with human hearts prior to the operation to insure its success.
   - If there is a choice between a baboon and a human life, the human life is more valuable.
   - Such operations can greatly improve our knowledge of such areas of medical research as immunology.

Against

- Such transplants are improper, being unfair both to the baboon donor and the human recipient.
- It is impossible to know whether the transplanted baboon heart will grow and develop as a human heart would as the person grows.
- It is wrong to breed baboons and sacrifice them solely to save human lives.
- We do not know the long-term psychological effects on a human recipient of a baboon heart.

8. Ask these follow-up questions of all board members:
   - How would their decision have been changed, if at all, if:
     —the organ to be transplanted had been a kidney instead of a heart? A liver?
     —this had been a human to human transplant?
     —the donor had been a chimpanzee instead of a baboon, closer evolutionarily to humans, with less of a chance of rejection than a baboon heart?
     —the infant had been female instead of male?
     —the parents had had a different socioeconomic background and marital history?
   - At present, parental consent is required for this type of transplantation. Should it be? Why or why not?

Answers to Questions on Student Worksheet

1. and 2. Answers will vary.
3. and 4. Answers will vary; refer to #8 under "Teaching the Activity" for examples of arguments for and against transplantation.

Follow-up Activities

- Have students read news articles about actual cases such as "Baby Fae" involving baboon to human transplants and human to human transplants.
- Repeat this activity, assuming that:
  —a human is the donor;
  —a chimpanzee is the donor;
  —a kidney instead of a heart is donated;
  —an adult instead of an infant received the donated organ.
- Bring in a surgeon or immunologist to speak about medical problems associated with transplantation and how these are being overcome.
- Have students do some research on the action of cyclosporin, a drug which helps prevent tissue rejection without paralyzing the entire immune system.
- Conduct a panel discussion of transplantation with members of the community. Be sure to prepare questions in advance. Individuals you might ask include:
  —persons who have received transplants
  —family members of persons who have received transplants
  —doctors and nurses (choose those who have participated in transplant operations, if possible)
  —counselors (of people receiving transplants, if possible)

—clergy members

—philosophy professors

Evaluation

Method of Evaluation	Goal(s) Tested
1. Participation in the simulation	1,2,3
2. Completion of student worksheets	1
3. Participation in class discussion	1,2,3
4. Questions 1 and 2	1
5. Question 3	2,3

Question 1:

List three arguments in favor of and three opposing baboon to human transplants.

Question 2:

Should baboon to human transplants be permitted? If so, under what circumstances? Support your viewpoint with sound arguments.

Question 3:

Describe one possible medical complication and one potential psychological problem for the human recipient of a baboon heart transplant.

References

1. Adler, J., Huck, J., and McAleney, P., "Baby Fae's Heart Gives Out", *Newsweek*, November 26, 1984, p. 94.

2. Clark, M., Huck, J., Raine, G., Sandza, R., Gosnell, M., and Hager, M., "A Breakthrough Transplant?", *Newsweek*, November 12, 1984, pp. 114–118.

SHOULD BABOON TO HUMAN TRANSPLANTS BE PERMITTED?
(BACKGROUND INFORMATION)

You are going to simulate a hospital bioethics review board procedure to decide whether a baboon to human heart transplant is to be permitted. Although such transplants have been attempted, to date none has been successful. Some members of your class will present the case; others will make the decision. The case and roles are described briefly below. Your teacher will provide further directions.

Case:

A young male infant has been born with a congenital heart deformity. If drastic measures are not taken soon, the infant will almost surely die. An experienced heart surgeon of the major hospital in which the child was born considers a baboon heart transplant to be the best solution.

Roles:

presenters:

1. the surgeon who wants to do the transplant
2. parents of the child who would receive the transplant (The persons choosing these roles may decide what their socioeconomic background, occupations, marital history, and beliefs are. They may or may not agree with the views of the surgeon. At present, parental consent is required for transplants of this type. For purposes of this activity, however, assume that it is not required.)
3. a representative of the American Society for the Prevention of Cruelty to Animals

board members:

1. doctors (up to three)
2. nurses (up to three)
3. administrators of the hospital in which the transplant will occur (two)
4. clergy members (at least two, each from a different denomination; if you choose this role, the denomination will be your choice)
5. a university philosophy professor who teaches medical ethics
6. nonmedical members of the community (at least two; people choosing these roles may decide their occupations, socioeconomic status, and beliefs.)
7. a lawyer

Name ——————————————————————— Date ———————————

SHOULD BABOON TO HUMAN TRANSPLANTS BE PERMITTED?
PREPARATION SHEET

1. What role have you chosen? ———————————————————————

2. Is this person for or against transplantation? Explain.

   ———

   ———

3. List arguments this person will cite in support of his or her viewpoint:

   a. ————————————————————————————

   ————————————————————————————

   b. ————————————————————————————

   ————————————————————————————

   c. ————————————————————————————

   ————————————————————————————

   d. ————————————————————————————

   ————————————————————————————

   e. ————————————————————————————

   ————————————————————————————

4. List counterarguments other members of the board or the presenters might make. For each, describe briefly how you will address each one:

   a. ——————————————————   ——————————————————

   b. ——————————————————   ——————————————————

   c. ——————————————————   ——————————————————

   d. ——————————————————   ——————————————————

   e. ——————————————————   ——————————————————

ACTIVITY 23

SHOULD TRANSMISSION OF THE AIDS VIRUS BE A CRIMINAL OFFENSE?

Goals
After completing this activity, students will:

1. understand how AIDS is transmitted;
2. be familiar with problems associated with controlling the spread of the AIDS virus; and
3. be aware of some ethical and medical questions raised by the AIDS epidemic.

Synopsis
Students will write a law establishing guidelines for dealing with persons who knowingly transmit the AIDS virus to others, or explain in detail why such a law should not be written.

Relevant Topics
microbiology
AIDS education
communicable diseases
bioethics
anatomy and physiology

Age/Ability Levels
grades 10–12, average to gifted

Entry Skills and Knowledge
Before participating in this activity, students should:

1. be able to state the cause and list symptoms of AIDS; and
2. be able to explain the dangers of transmitting the disease to another individual.

Materials

Student worksheets: "Should Transmission of the AIDS Virus Be a Criminal Offense" and "Facts about AIDS"

DIRECTIONS FOR TEACHERS

Preparation
1. Give your principal a copy of this activity several weeks in advance to be sure there are no problems with using it.
2. Obtain necessary school approval and parental permission.
3. Determine, if possible, whether there are any students attending your school who have AIDS; you will need to be sensitive to their feelings as you conduct the activity.
4. Have students read a current article about AIDS or the "Facts About AIDS" sheet.

NOTE: In addition to the fact sheet and references listed in this book, information about AIDS is available from:

The National AIDS Information Clearinghouse (NAIC)
P.O. Box 6003
Rockville, MD 20850
Phone: (1-301-762-5111)

and:

The National AIDS Hotline
1-800-342-AIDS OR 1-800-344-SIDA (for Spanish-speaking callers)

Teaching the Activity

1. Briefly explain the assignment. Distribute worksheets, and give students time to read them.

2. Students will probably have many misconceptions about the spread of AIDS. Take time to answer their questions before they begin working. Here is some information that might be helpful:

   Ways to prevent the spread of the virus for each of the methods of transmission include:

   Intimate sexual contact: Use condoms, educate people about the dangers, abstain from sex, have only one sexual partner, ask the partner prior to sex if he or she is a carrier of AIDS.

   Blood transfusions: Screen blood before it is used, ask high risk groups not to donate blood, question people carefully before they give blood, educate people about the dangers of giving blood if they are carrying the AIDS virus.

   Sharing needles during intravenous drug use: Do not use drugs, do not share needles, educate people about the dangers.

   Transmission from mother to child before or during birth: screen the blood of women before they become pregnant, educate people about the dangers.

3. Allow time in class to complete worksheets. It is your decision as to whether work is done individually or in small groups. Give help as needed, but most ideas should come from students.

4. When all have finished, ask volunteers to share their guidelines and reasons for them. Record these on the board. Ask those who had strong feelings against writing a law to share their thoughts.

5. If students fail to bring up the following points and questions during discussion, ask for their reactions to them:

   • Is it right to enact a law which limits personal freedom, even if the purpose is to prevent the spread of a fatal disease such as AIDS? Does the end justify the means in this situation?

   • Which persons in each of the scenarios below, if any, should be considered criminals? Should their punishments be the same? Explain.

     —A homosexual man knows he is carrying the AIDS virus. He continues to have sex with many different partners, without informing them of the risk.

     —A pregnant woman unknowingly contracts AIDS from her husband, who had an extramarital affair with another woman. She passes the virus on to her unborn child.

—A pregnant woman unknowingly contracts the AIDS virus from a man during an extramarital affair. She passes it on to her unborn child.

—A heterosexual woman knows she is carrying the AIDS virus. She continues to have sex with many different partners, without informing them of the risk.

—A heterosexual man knows he is carrying the AIDS virus. He continues to have sex with many different partners, without informing them of the risk.

—A woman unknowingly received the AIDS virus during a transfusion prior to the 1985 blood screening program. She marries and becomes pregnant, passing the virus on to her child and to her husband.

—Two intravenous drug users routinely share the same needle. One knows he carries the AIDS virus but has never told the other person.

—A heterosexual woman learns that she is carrying the AIDS virus. She thinks she received it from a man she knew before she was married. She fails to tell her husband about the other man, or about the risk to him.

—A man does not know he is carrying the AIDS virus and gives blood routinely to hospitals.

• Who are the "victims" in each of the scenarios above? Should they be informed that they have been exposed to the AIDS virus? Explain.

• Would the legal guidelines you wrote be adequate in dealing with each of the above scenarios? If not, how should your guidelines be changed?

6. Ask students whether or not they think any of the laws they have just written should be enacted. If so, have them perform the first follow-up activity.

7. Additional questions for discussion include these:

• Should all persons carrying the AIDS virus be quarantined? If so, how would it be done?

• Should all people be screened routinely to determine whether they are carriers of the AIDS virus? If so, to which of the individuals or organizations below, if any, should test results be made available?

—the persons being screened

—their employers

—their insurance companies

—their hospitals

—their sexual partners

—their school officials, if they are students

What might each of the above do with the information once it was received? What would be the consequences if the test gave false positive or false negative results, as sometimes happens?

• It is often said that children are the "innocent" victims of AIDS. If this is true, does this mean that all other victims are "guilty"? In what ways? Define "innocence" and "guilt" as they are used in this context.

Answers to Questions on Student Worksheets

1. Answers will vary; refer to #2 under "Teaching the Activity" above for answers.

2. through 8. Answers will vary.

9. Answers will vary; blood of the individuals could be screened.

10. and 11. Answers will vary.

Laws Students Write will vary; they should be based on answers to questions 1–11.

Follow-up Activities

- Have students listen to a representative from the city, state, or federal government speak about how to present ideas concerning AIDS to those in a position to make and change laws. Students may then want to write letters, make phone calls, or visit lawmakers in person to discuss laws dealing with AIDS.

- In a hypothetical situation, a television evangelist has stated that it should be illegal for all homosexuals to have intercourse because they may transmit AIDS. Have students list arguments for and against such a proposal.

- If a law such as that described in the follow-up activity above were enacted, should it also apply to:

  —homosexual affairs in which the sexual partners give written voluntary and knowing consent?

  —bisexual persons married to heterosexuals?

  —anyone, heterosexual or homosexual, who has been shown to carry the AIDS virus?

- Have students find out what methods to prevent the spread of the AIDS virus are being enacted in their community.

- Have students keep up with current legal precedents concerning AIDS.

- Ask students to explain why antibiotics cannot be used to cure AIDS.

- Assign research on current treatments for AIDS, their modes of action, availability, side effects, and effectiveness.

- Have students propose a cure for AIDS which demonstrates an understanding of the biochemical basis of RNA virus reproduction.

- Should dentists and dental technicians refuse to treat AIDS patients? Some fear that the blood of the patient will come in contact with that of the person treating him, allowing easy transmission of the virus. Ask students to:

  —List arguments in favor of and opposed to their refusing to treat these patients.

  —List precautions dentists and technicians might take as they treat AIDS patients to minimize the risk of contracting the virus.

- In a hypothetical situation, a group of nurses works in a hospital in which their employers will not let them wear gloves to treat AIDS patients. They are drafting a letter of protest to send to the hospital administrators. Have students cite ethical and biological arguments supporting and refuting their position.

- Bel Glade, Florida is a community in which there has been controversy regarding AIDS. Some have thought that mosquitoes have been transmitting the AIDS virus among humans. Have students research the case and present evidence in favor of and against the idea that mosquitoes might be carriers of AIDS.

Evaluation

Method of Evaluation	Goal(s) Tested
1. Participation in class discussions	1,2,3
2. Completion of student worksheets	1,2,3
3. Question 1	1,2,3

Question #1

List four ways the AIDS virus is transmitted. Beside each, cite

- two ways the spread of the virus by this means might be controlled; and
- an ethical issue that would be raised by making each of these practices a law.

References

1. Adler, J., Greenberg, N. F., Hager, M., McKillop, P., and Namuth, T., "The AIDS Conflict," *Newsweek*, September 23, 1985, pp. 18–24.

2. Bastian, B., "AIDS: Education Against Fear," *Carolina Tips*, February 1, 1986 (a publication of Carolina Biological Supply Company; Burlington, N.C., 27215).

3. Hurley, J., "I'm Dying—AIDS Is Your Problem Now," *Newsweek*, August 10, 1987, pp. 38–39.

4. Koop, C. E., *Surgeon General's Report on Acquired Immune Deficiency Syndrome*, October 22, 1986 (available from the U.S. Department of Health and Human Services).

5. Seligmen, J., Gosnell, M., "AIDS: Myth and Reality," *Newsweek*, September 23, 1985, pp. 20–21.

6. "What Science Knows About AIDS: A Single Topic Issue," *Scientific American 259*(4), October, 1988.

SHOULD TRANSMISSION OF THE AIDS VIRUS BE A CRIMINAL OFFENSE?

Your assignment is to write a law establishing guidelines for dealing with people who knowingly transmit the AIDS virus to others. If, however, after you have completed the following questions, you strongly believe such a law should not be written, explain these beliefs in detail instead of writing a law. Before writing your law, complete the following questions. Use a separate sheet of paper for your answers.

1. For each mode of transmission, list ways to prevent the spread of the AIDS virus: a) intimate sexual contact, b) blood transfusions, c) sharing needles during intravenous drug use, d) transmission from mother to child before or during birth.

2. Should persons transmitting the disease to someone else by each of the four modes listed above be dealt with the same way under the law? If not, what differences should there be and why?

3. Should those who knowingly transmit the virus be dealt with in the same way as those who transmit it unknowingly? If not, how will it be determined who was a knowing transmitter, and who was an unknowing one?

4. Which of the methods you described in question 1, if any, should be mandated by law?

5. What kinds of punishment will prevent the spread of the virus and at the same time protect a) the rights of the "victims" and b) the rights of the "criminals"?

6. For your answers to questions 4 and 5, list potential problems with enforcing these laws.

7. Should any law of this type include a hierarchy of penalties, becoming more severe as the number of offenses increases? If so, what kinds of penalties would be appropriate?

8. If the person receiving the virus (the "victim") signs a written statement that he or she is aware of the risks, should the act of transmission then be legal?

9. How will persons spreading the AIDS virus be identified?

10. Should AIDS "victims" be informed that they may have received the virus? If so, how?

11. Should a law dealing with the spread of AIDS be enacted by individual states or by the federal government? Explain.

Writing Your Law:

Write your law, using the answers to the questions above as the basis. If, however, you now strongly believe such a law should not be written, explain these beliefs in detail instead of writing the law.

FACTS ABOUT AIDS:*

Most scientists agree that:

- the AIDS virus can only be spread through intimate sexual contact, blood transfusions, sharing a contaminated needle or syringe during intravenous drug use, or from an infected mother to her child before or during birth; and
- the virus must be transmitted to the blood through an opening in the skin, rectum, anus, penis, or vagina.

The HTLV-III virus (human T-lymphotrophic virus type III) has been found in the blood, semen, saliva, tears, cerebrospinal fluid, and breastmilk of AIDS PATIENTS. *Note:* The HTLV-III virus is also known as "HIV" (human immunodeficiency virus).

A person can acquire the AIDS virus during sexual contact with an infected person's blood, semen, and possibly vaginal secretions. So far, there is no evidence that the virus can be spread through casual contact, including social kissing, hugging, shaking hands, crying, coughing, and sneezing.

Transmission can occur in these ways:

- during anal intercourse (homosexual or heterosexual)
- during vaginal intercourse (from either partner to the other)
- before birth—either crossing the placenta from mother to fetus, or during birth
- receiving blood during a transfusion (Since March 1985 blood in hospitals has been screened routinely for the presence of the antibody to the AIDS virus. Contaminated blood is discarded, making most transfused blood safe. However, a newly infected person may donate blood before the antibodies have had a chance to form; this is estimated to occur less than once in 100,000 donations.)
- sharing a needle during intravenous drug use that is contaminated with blood from someone carrying the AIDS virus.

A person CANNOT get AIDS from giving blood, because sterile needles are used to withdraw blood only once, then discarded.

It is highly unlikely that a person can get AIDS from toilet seats or doorknobs, because the virus dies quickly when exposed to air. Also, most experts agree that an open cut in the skin through which the virus must enter would be necessary.

HTLV-III is an RNA virus, using reverse transcriptase to copy itself into DNA in the host cell.

HTLV-III attacks T-lymphocyte cells of the human immune system, which are "helper" cells necessary for the production of antibodies.

The virus can be killed by 10% household bleach solution, 1% glutaraldehyde solution, 70% ethanol, exposure to heat, or exposure to air.

Patients with the AIDS virus die from secondary infections because their immune systems are weakened. Two illnesses which commonly cause death in this way are Kaposi's sarcoma, a rare form of cancer, and *Pneumocystis carinii* pneumonia.

Not all people who carry the AIDS virus develop the disease.

---

* This information is summarized from Bastian, B., "AIDS Education Against Fear," *California Tips,* Feb. 1, 1986, © Carolina Biological Supply Company, Burlington, N.C. 27215, used by permission, and Koop, C.E., *Surgeon General's Report on Acquired Immune Deficiency Syndrome,* Oct. 22, 1986, public domain material available from the U.S. Department of Health and Human Services.

section 7

Biology Learning Games

Playing science games in the classroom is a great way to motivate students. I have heard students comment that learning biology in the context of a game seemed like fun, not work.

Moreover, when students compete in teams, each team member has a necessary role and is valued. Group work in team competition is a way to have students of differing ability and socioeconomic levels work together, perhaps for the first time. In this way students can learn mutual respect and cooperation.

Since a game provides students with a definite goal, and since they enjoy games, classes are usually easy to manage.

Characteristics of the three games in this section include the following:

1. All require that students use their knowledge of biology either to solve hypothetical problems (as in "What If There Were No Fungi?") or to draw relationships between complex biological concepts (as in "What Are Relationships Between Cell Structures and Photosynthesis Processes?" and "What Are Relationships Between Cell Structures and Cellular Respiration Processes?"). Both types of endeavors—solving hypothetical problems and drawing relationships—are important in scientific work and encourage active thinking.

2. The games involve interaction and discussion with other students and with the teacher, so that students learn not only from you, but from each other, as well.

3. All of the games have an object, with rules and procedures. The "true" object is to help students better understand biology, but the stated game object is to have teams accumulate points toward a grade. It will be your decision whether or not to reward the winning team in another way, as well—with bonus points, for example, or a pizza.

4. The games are inexpensive, since most of what you need to make game materials is contained here. All you need, in addition to what is in this section, is heavy white paper, construction paper, envelopes, a photocopy machine, and a laminating machine (optional).

The games entitled "What Are the Relationships Between Cell Structures and Photosynthesis Processes?" and "What Are the Relationships Between Cell Structures and Cellular Respiration Processes?" have exactly the same rules and procedures. The terms and cell structures referred to in the games differ, obviously.

How Should You Use These Activities?

The games included here can be used as they are written, or you can modify them for other subject areas, as described in some of the follow-up sections. Another way to use games in your classes is to have students create their own biological games, incorporating characteristics described above. Students may wish to model their games on commercial ones they have played or seen on television, or create their own rules.

You can create games easily using the rules and procedures of the "relationships" games here. Any science content area can be a source of terms from which students draw relationships.

Suggestions for integrating these games into a standard high school biology course include the following:

1. All three games make excellent reviews for tests in each of their subject areas, since they require students to "pull together" information they have already studied. This is particularly true of "What Are Relationships Between Cell Structures and Photosynthesis Processes?" and "What Are Relationships Between Cell Structures and Cellular Respiration Processes?"

2. "What If There Were No Fungi?" is appropriate as part of a unit on microbiology or ecology. An interesting strategy is to have students list the consequences of removing fungi from earth before they study fungi in your class, and afterwards. Comparing the two sets of answers will let you know how much the students have learned about the importance of fungi.

ACTIVITY 24 _____

WHAT ARE RELATIONSHIPS BETWEEN CELL STRUCTURES AND CELLULAR RESPIRATION PROCESSES?

Goals

After participating in this activity, students will:

1. better understand cellular respiration;
2. be able to identify where in the cell different phases of cellular respiration occur; and
3. be able to state relationships between various molecules and processes associated with respiration.

Synopsis

Students will play a game in which they state relationships between parts of the cell, processes, and molecules associated with cellular respiration.

Relevant Topics

cellular structure
cellular respiration
human respiratory system

Age/Ability Levels

grades 9–12, average to gifted

Entry Skills and Knowledge

Before participating in this activity, students should be able to:

1. identify the parts of the cell in which the various phases of cellular respiration occur; and
2. describe aerobic and anaerobic respiration.

Materials

game cards
construction paper
envelopes
(optional) evaluation questions worksheet

DIRECTIONS FOR TEACHERS

Preparation

1. Make the game cards by photocopying the card sheets onto heavy paper and cutting out individual cards along solid lines. Cards will last longer if they are laminated.
2. Pull out cards with terms and cell structures your class has not covered. In this way, you can adjust the content and difficulty of the game to student ability level.

3. Decide on a system of grading. Below are some suggestions for ways of converting correct answers into grades:

—Have answers comprise a daily grade by, for example, making each correct answer worth 10 points.

—Have answers count as extra credit points added to a test grade by, for example, making each correct answer worth ¼, ½, or 1 point.

4. Announce the game to students at least one day in advance so that they can study for competition.

Teaching the Activity

1. Two types of cards will be used in the game—one type lists cell structures, such as the cytoplasm; the other type lists respiration terms, such as glycolysis. Shuffle ALL cards into one deck and turn them face down.

   NOTE: Shuffling all cards allows for the possibility of drawing relationships between two terms, or between two structures. If you think this is too difficult for your students, keep the "structures" and "terms" cards separate as students draw them.

2. Explain the game procedures to students:
   a. One student will draw a card from the deck in your hand and read it aloud to the entire class.
   b. Another student will draw a second card and read it aloud.
   c. A third student will volunteer to state a relationship between the two cards.
   d. You will give a token, such as a small square of red construction paper, to the student if the answer is correct. If it is incorrect, call for another volunteer. Be sure all students understand why answers are wrong. In some cases, the correct answer will be "no relationship exists." In this situation, the student should be given a token for this answer. Students are not permitted to use notes and books as they play the game. Examples of correct answers are given below.
   e. Repeat the process as long as time permits, for at least 15 minutes. Students should keep a record on their own paper of relationships they miss or do not understand.
   f. At the end of the game, have each student who received tokens place them in an envelope with his or her name on it and then pass the envelope to you. Record the number of points each student received.

   NOTE: If you are afraid students might cheat by making their own tokens, code the tokens that you distribute in advance with a numbering system and keep track of which numbers are handed out.

Game Answers

I. *Relationships Between Cell Structures and Respiration Terms*

Some possible answers are listed below. The cell structure is given first, the term second, and the relationship third.

cytoplasm—glycolysis—Glycolysis occurs in the cytoplasm.

mitochondrial matrix—electron transport chain—The Krebs cycle, located in the mito-chondrial matrix, feeds electrons into the electron transport chain via NADH and $FADH_2$.

mitochondrial outer membrane—oxidative phosphorylation—The mitochondrial outer membrane helps maintain a proton gradient needed for oxidative phosphorylation to occur.

cytoplasm—oxidative phosphorylation—Pyruvate, made in the cytoplasm, is involved in the Krebs cycle, which participates in oxidative phosphorylation.

cytoplasm—FAD—No relationship (FAD participates in reactions occurring in the mitochondrial matrix, so there is no DIRECT relationship; however, FAD building blocks are made in the cytoplasm).

cristae—oxygen—Oxygen is the ultimate electron acceptor for the electron transport chain, located in the cristae.

cristae—carbon dioxide—Carbon dioxide is a by-product of the Krebs cycle, which feeds electrons into the electron transport chain, located in the cristae.

II. *Relationships Between Respiration Terms*

ATP—NADH—NADH is an electron carrier in the processes used to make ATP.

Krebs cycle—glycolysis—Both processes make ATP; both involve reduction-oxidation reactions and electron carriers; pyruvate from glycolysis is involved in the Krebs cycle.

electron transport chain—pH—A pH gradient, thought to provide energy for oxidative phosphorylation, is thought to be maintained in part by the electron transport chain.

oxygen—water—Oxygen is the ultimate electron acceptor for the electron transport chain; as it receives electrons, water is produced as a by-product.

carbon dioxide—oxidative phosphorylation—Carbon dioxide is a by-product of oxidative phosphorylation.

FAD—NAD—Both of these are electron carriers, participating in ATP production.

glucose—energy—The energy from glucose is used to make ATP, the "energy currency" of living cells.

Follow-up Activities

- Have students point out the areas on commercial models of a cell and a mitochondrion in which the various phases of respiration occur.

- Instruct students to make a model of a cell or mitochondrion using materials such as plaster of paris, food, wire, or plastic toys.

- Display electron micrographs of mitochondria in cross section. If possible, show micrographs from various types of cells and different types of organisms—healthy and diseased. Ask these questions about the micrographs:

—Where on the micrograph do the various phases of respiration occur?

—How do mitochondria differ between cells of various organisms? Explain.

—How do they differ between various cells of the same organism? Explain.

—Do the mitochondria of normal cells appear the same and have the same function as those of diseased cells? If not, how do you explain the differences?

Evaluation

Method of Evaluation	Goal(s) Tested
1. Participation in the game	1,2,3
2. Evaluation questions worksheet	1

Evaluation question 1 answers:

a. 3   b. 108   c. 0

Evaluation question 2 answers:

a. 3   b. 2   c. 1   d. 2   e. 3   f. 1

RELATIONSHIPS BETWEEN CELL STRUCTURES
AND CELLULAR PROCESSES
Evaluation Questions

Question 1

A cell undergoing anaerobic respiration produces a net of six ATP molecules from x glucose molecules. From this information, answer the questions below in the spaces provided:

_____a. How many glucose molecules were used?

_____b. If the cell described were aerobic instead of anaerobic, how many ATP molecules (net) would have been produced?

_____c. If all of the cytoplasmic enzymes of the anaerobic cell were inactivated, how many ATP molecules (net) would be produced?

Question 2

Indicate which of the following is a characteristic of lactic acid fermentation, alcohol fermentation, or both, by writing the appropriate numbers in the spaces provided. Use this code:

1 = lactic acid fermentation
2 = alcohol fermentation
3 = both

_____a. Hydrogen atoms are transferred.

_____b. Certain yeasts and bacteria perform the process(es).

_____c. The purpose is to help relieve the "oxygen debt" after strenuous exercise.

_____d. Ethanol is produced.

_____e. NAD+ is regenerated.

_____f. The human muscle cells are where it/they occur(s).

CELL STRUCTURES GAME CARD SHEET

cytoplasm	mitochondrial outer membrane
mitochondrion	cristae
mitochondrial inner membrane	mitochondrial intermembrane space
mitochondrial matrix	

proton gradient	oxidative phosphorylation
pH	substrate level phosphorylation
alcohol fermentation	FAD
lactic acid production	FADH$_2$
aerobic respiration	ADP

RESPIRATION TERMS GAME CARD SHEET 2

anaerobic respiration	ATP
electron transport chain	glucose
oxidation-reduction reactions	NAD
electron transport	NADH
cardon dioxide	oxygen

RESPIRATION TERMS GAME CARD SHEET 3

energy	water
electrons	citric acid
pyruvate	glycolysis
Krebs cycle	cytochromes
enzymes	

ACTIVITY 25

WHAT ARE RELATIONSHIPS BETWEEN CELL STRUCTURES AND PHOTOSYNTHESIS PROCESSES?

Goals

After participating in this activity, students will:

1. better understand photosynthesis;
2. be able to identify areas of the cell in which various phases of photosynthesis occur; and
3. be able to state relationships between molecules and processes associated with photosynthesis.

Synopsis

Students will play a game in which they state relationships between parts of the cell, processes, and molecules associated with photosynthesis.

Relevant Topics

cellular structure
photosynthesis
plant anatomy

Age/Ability Levels

grades 9–12, average to gifted

Entry Skills and Knowledge

Before participating in this activity, students should be able to:

1. identify the parts of the cell in which the various phases of photosynthesis occur; and
2. describe the process and purpose of photosynthesis.

Materials

game cards
construction paper
envelopes
(optional) evaluation questions worksheet

DIRECTIONS FOR TEACHERS

Follow the same preparation and teaching steps described in the preceding activity ("What Are Relationships Between Cell Structures and Cellular Respiration Processes?"). See page XXX.

Game Answers

I. Relationships Between Cell Structures and Photosynthesis Processes Terms

Some possible answers are listed below. The cell structure is given first, the term second, and the relationship third.

*thylakoid (or grana)—light reactions—*The light reactions of photosynthesis occur in the thylakoids.

*stroma—dark reactions (or Calvin cycle)—*The dark reactions (or Calvin cycle) occur in the stroma.

*chloroplast—chlorophyll—*Chlorophyll, located in chloroplasts, is responsible for their green color.

*cell wall—sugars—*Sugars, which are products of photosynthesis, are used to build cell wall polysaccharides.

*cytoplasm—P680—*No direct relationship, although P680 does reside in the chloroplast, which is suspended in the cytoplasm.

*thylakoid interior—electron transport—*Electron transport helps maintain the proton reserve in the thylakoid interior, which is necessary for ATP production.

*thylakoid—NADP—*NADP is reduced in the thylakoids to NADPH to be used in the dark reactions.

II. *Relationships Between Photosynthesis Terms*

*ATP—NADPH—*Both molecules are products of the light reactions and are used in the dark reactions of photosynthesis.

*enzymes—sunlight—*The enzymes of the Calvin cycle help transform energy from sunlight into sugars.

*ADP—sugars—*ADP is a by-product of the process of converting smaller molecules to sugars in the Calvin Cycle.

*Photosystem I—noncyclic photophosphorylation—*Photosystems I and II are involved in the electron flow necessary for noncyclic photophosphorylation.

*water—electrons—*Electrons, released as water is broken down, are transferred to P680.

*carbon dioxide—oxygen—*These are both gases the plant exchanges with the atmosphere; oxygen is a by-product of photosynthesis; CO_2 is used in photosynthesis.

*dark reactions—Calvin cycle—*The Calvin cycle is involved in the dark reactions of photosynthesis.

Follow-up Activities

- Using commercial models of cells and chloroplasts, have students identify the areas in which the different phases of photosynthesis occur.
- Assign students to make models of chloroplasts using materials such as plastics, wire, and food.
- Exhibit electron micrographs of chloroplasts. Have students identify the regions on the micrographs in which the different phases of photosynthesis occur.
- Have students view prepared slides of chloroplasts under the microscope.
- A theory of the origin of chloroplasts and mitochondria is that they were once free-living and became symbiotic with eukaryotic cells. Assign students to research this topic, 1) citing evidence supporting the idea, and 2) indicating which regions of the chloroplast are coded for by chloroplast DNA and what areas by nuclear DNA.
- Have students observe aquatic and terrestrial plants, and 1) cite evidence that each type of plant is undergoing photosynthesis and 2) design experiments to provide more evidence that photosynthesis is occurring.

Evaluation

Method of Evaluation	*Goal(s) Tested*
1. Participation in game	1,2,3
2. Evaluation Questions worksheet	1

Evaluation question 1 answers:

a. NC   b. C   c. C   d. NC

Evaluation question 2 answers:

a. Y   b. N   c. Y   d. Y

Name ———————————————— Date ————————————

RELATIONSHIPS BETWEEN CELL STRUCTURES
AND PROCESSES OF PHOTOSYNTHESIS

Evaluation Questions

Question 1

Experiments are conducted on green plant cells at various times to determine whether they are undergoing cyclic or noncyclic photophosphorylation. From the results of each experiment described below, indicate whether the cells are undergoing cyclic or noncyclic photophosphorylation, using the following code:

C = cyclic
NC = noncyclic

—————————a. A great deal of NADPH is made.

—————————b. Water is not split, so no oxygen is released.

—————————c. The pigment P700 serves as the original electron donor as well as the final electron acceptor.

—————————d. All of the components of Photosystems I and II are involved.

Question 2

A new herbicide has been developed that blocks electron flow in the light reactions of photosynthesis. Indicate whether each possible consequence of spraying the herbicide on a green plant would or would not occur by writing a letter of the following code in the spaces provided.

Y = Yes, it would occur.
N = No, it would not occur.

—————————a. ATP production would cease.

—————————b. The plant would live indefinitely.

—————————c. NADPH would not be produced.

—————————d. New glucose could not be synthesized.

CELL STRUCTURES GAME CARD SHEET

cytoplasm	chloroplast
chloroplast stroma	chloroplast grana
thylakoids	intergranal lamellae
cell wall	thylakoid membrane
thylakoid interior	

PHOTOSYNTHESIS TERMS GAME CARD SHEET 1

proton gradient	cyclic photophosphorylation
pH	noncyclic photophosphorylation
light reactions	dark reactions
chlorophyll	sugars

Z-scheme	ATP
sunlight	ADP
electron transport chain	Photosystem I
oxidation-reduction reactions	NADP

PHOTOSYNTHESIS TERMS GAME CARD SHEET 3

electron transport	NADPH
carbon dioxide	oxygen
Photosystem II	water
electrons	phosphoglyceraldehyde

photophosphorylation	P680
Calvin cycle	cytochromes
enzymes	P700
photosynthesis	

ACTIVITY 26 _____
WHAT IF THERE WERE NO FUNGI?

Goals
After participating in this activity, students will:

1. understand some of the consequences of removing fungi from earth; and
2. be more familiar with the positive and negative values of fungi.

Synopsis
Students will compete in groups as they list the consequences of destroying all the fungi on earth.

Relevant Topics
microbiology
ecological balance
fungi
decomposers in ecosystems

Age/Ability Levels
grades 9–12, most ability levels

Entry Skills and Knowledge
Before participating in this activity, students should be able to:

1. list different types of fungi; and
2. list some of the benefits and dangers of fungi to plants and animals

Materials
Student worksheets: "What If There Were No Fungi? Game Directions" and "What If There Were No Fungi? Answer Sheet"
carbon paper or backs of old ditto masters
teacher record sheets (included)

DIRECTIONS FOR TEACHERS

Preparation
1. Obtain carbon paper or backs of old ditto masters.
2. Decide on a grading system for the game and a way to reward the winning team; suggestions for ways to count points include:
   - as a daily grade, in which each correct consequence is worth 10 to 15 points; or
   - as extra credit points added to a test grade (each correct consequence could be worth ¼ or ½ point).

   The team members with the highest number of points could receive bonus points (added to the grade earned for the game) or a pizza.
3. Describe the game to students at least one day in advance so that they can study.
4. Photocopy student directions and answer sheets.

Teaching the Activity

1. Distribute student directions, answer sheets, and carbon paper, and allow students time to read the directions. Explain the grading system, and answer any questions.

2. Have students work in teams of three or four members each and assign each group a different number. Record this information on the Teacher Record Sheet.

3. Be sure students are seated so that they can easily discuss their ideas with each other. Suggestions for seating include pulling desks into a circle, seating students on either side of a lab table, or seating students on the floor.

4. Allow groups at least 10 minutes to compile their lists.

5. While students are answering question 4 on their worksheet, mark their answers as correct or incorrect by making a check in the appropriate box of the answer sheet. Be sure to mark carbon copies, as well. Checking may be done overnight.

6. On your record sheet, record the total number of points for each team.

7. Return sheets to team members. Allow them to use their notes and books to change incorrect responses. Some correct consequences include these:

   - Wine and beer production using brewer's yeast would be impossible; people who make these products would go out of business.

   - Yeast breads and rolls would not exist; a leavening agent such as baking powder would have to be used instead of yeast, or unleavened bread could be eaten. People selling yeast dough products would be required either to change their recipes, or go out of business.

   - Dead plants and animals would be decomposed only by bacteria, since bacteria and fungi are the two major decomposers on earth. Since many fungi can grow under conditions hostile to most bacteria, dead plant and animal matter would tend to accumulate. This would mean that soils would be less rich than those of today and that crops not provided with extra fertilizer would suffer.

   - The following would not exist:

     athlete's foot

     rotting of fruits and vegetables

     fungal plant diseases such as potato blight

     mold on bread

     ringworm

     mold growths on clothes, shoes, or books

   - The following would also not exist:

     Brie, Camembert, and Roquefort cheese

     mushrooms (edible or poisonous)

     interesting fungi for students to observe

     penicillin (except that produced synthetically)

     certain types of soy sauces

     truffles

   - Lichens would cease to exist. Tundra animals which eat lichens would be forced to find other sources of food.

   - Symbiotic mycorrhiza, which help certain trees grow taller, would cease to exist.

8. Share consequences as a group. Have students explain why they think fungi should or should not be destroyed.

Answers to Questions on Student Worksheet

Table 26–1 Answers will vary; refer to #7 under "Teaching the Activity" for some correct responses.

Follow-Up Activities

- Have students read about the process of wine production.
- Have students prepare a menu containing at least three different fungi (such as bread, mushrooms, Brie).
- Lead a walk through the woods after it has rained. Using a field manual, identify the fungi you see, making a record of their numbers and varieties.
- Play a game such as this one that you create around other topics in biology, having students answer these questions:

  What if:

  —all chlorophyll on earth were destroyed?

  —all mitochondria in humans were destroyed?

  —the human body had only one kidney?

  —the human body had no liver (or heart, pancreas, white blood cells, etc.)?

  —all phytoplankton were removed from the oceans?

  —all whales were destroyed?

  —all bacteria and viruses on earth were destroyed?

  —humans could regenerate limbs?

  —humans could control the aging process?

You might want to omit the competition aspect of the activity for the last two questions above, and simply have a discussion about the answers.

Evaluation

Method of Evaluation	Goal(s) Tested
1. Completion of student worksheets	1,2
2. Participation in class discussion	1,2
3. Question 1	2

Question 1

List three benefits and three dangers of fungi to human beings.

Name _____ Date _____

WHAT IF THERE WERE NO FUNGI?
GAME DIRECTIONS

Pretend there has been a severe blight destroying a major corn crop in the United States. A group of citizens from several walks of life now feels that the solution is to release a newly synthesized chemical into the environment which has been shown to eradicate the fungus causing the blight. An environmental group argues that this chemical could destroy other fungi, as well, which are beneficial to humans. The citizens' group counters this argument by citing damage caused by fungi, such as spoilage to foods and mold growth on clothes. They believe that if all the fungi on earth were destroyed, our planet would be better off.

You are part of a scientific team assigned to predict the consequences of destroying all fungi on earth. The information you generate will be used to determine whether the plan to release the chemical should be implemented.

1. Work in teams as directed by your teacher.
2. Your team is to write a list of consequences, both positive and negative, of destroying all fungi on earth. Each correct consequence is worth a certain number of points toward a grade. The group with the highest number of points will receive a special reward, which your teacher will describe. All members of the team will receive the same number of points.
3. Write your list on the answer sheet your teacher gives you. One person in the group should be the recorder, making a carbon copy for each group member. Notes and books are not allowed during this exercise.
4. When you have finished your list, hand all copies to your teacher to be checked, and answer these questions:
   - What would you recommend to the citizens wanting to destroy the fungi?
   - Should they proceed with their plan? Explain.

Name _____ Date _____

WHAT IF THERE WERE NO FUNGI?
ANSWER SHEET

Team # _____

Names of team members:

Table 26–1. List of Consequences if All Fungi were Destroyed

Consequences	Correct	Incorrect
1.		
2.		
3.		
4.		
5.		
6.		
7.		
8.		
9.		
10.		

TEACHER RECORD SHEET

Team #	Names of Team Members	Total Points
1.	1. 2. 3. 4.	
2.	1. 2. 3. 4.	
3.	1. 2. 3. 4.	
4.	1. 2. 3. 4.	
5.	1. 2. 3. 4.	

TEACHER RECORD SHEET (continued)

Team #	Names of Team Members	Total Points
6.	1. 2. 3. 4.	
7.	1. 2. 3. 4.	
8.	1. 2. 3. 4.	
9.	1. 2. 3. 4.	